AF843669

# "WARWIRE"

## THE HISTORY OF OBSTACLE WIRE USE IN WARFARE

By DELBERT TREW

Revised Edition

Revised Edition
April, 2003



Printed in the United States of America

ISBN: 0-9659677-3-5

Address comments and requests to

Delbert Trew
Box A
Alanreed, TX 79002
(806) 779-3164
<trewblue@centramedia.net>

Brochures on books and prices sent on request

TABLE OF CONTENTS

Warwire History Text ........ page 5
Warwire Design ........ page 23
Warwire Tools ........ page 41
Warwire Specifications ........ page 79
Barrier Design & Installation ........ page 83
Manufacturing Warwire ........ page 97
Warwire Photography ........ page 109
Questions & Answers for collectors ........ page 123

THANKS FOR THE HELP OF THE FOLLOWING FRIENDS

Harold Hagemeier -wire identification and history
Jerry L. Janzen - bayonet identification and illustrations
Larry Greer - patent research and acquisition
Harold Ganshirt - history and research
Parolin Amedo - WWI photography
Mrs. Layne Myers & Pat - manufacturing history and photos
Brad Pinka - research and photo reproduction
Ann Williams - library research
Ruth Trew - editing, patience & encouragement

# INTRODUCTION

Why write a book about entanglement or obstacle wire, or as the military calls it, anti-intrusion wire? Others know it as concentration camp wire and modern war veterans refer to it as razor wire.

Historians and wire collectors have titled barbed wire as "The Devil's Rope." When barbed wire is used in warfare, the various designs are listed in the category of "Warwire."

Why would anyone read a book about this horrible device causing so much agony and suffering? The average reader shutters in fear upon seeing a photograph of a barbed wire barrier holding people prisoner.

Many photographs are designed to show the indignity and desolation of being confined behind a barbed wire barrier in hope the sight will be a graphic deterrent to crime and war. Some believe confining humans behind barbed wire or using it in war, is the ultimate insult in man's inhumanity to his fellow man.

But just a moment please, we have only referred to the use of barbed wire against humans. Its original use was intended to control livestock. The original inventors were farmers and stockmen.

At the time in 1874, there was a critical need for an economical device to control livestock and protect growing crops. The invention of barbed wire filled that need successfully.

It was so successful modern historical groups voted the barbed wire patent one of the most significant patents to emerge from The Industrial Revolution. When we study the inventions from this innovative period, barbed wire becomes a significant subject.

In the settlement of the West, barbed wire gave the settlers control of their land. This control resulted in approximately two-thirds of the United States being settled *after* the invention of barbed wire.

These facts should convince the reader the invention and use of barbed wire in non-war situations is important and necessary. But, back to the original question of why write a book about the use of barbed wire in warfare?

Research reveals the use of barbed wire in warfare drastically changed age-old military tactics, revised much of the equipment used in war, and required the training and equipping of special units to cope with the barriers.

In fact, the use of barbed wire in warfare had as much impact on war as it had on the rural economy of the private sector. And, anything causing this much change in the way things are ordinarily done is certainly worth writing about.

# "WARWIRE"

## BY DELBERT TREW

## CHAPTER 1. 1350-1850

One of the oldest forms of metal working is the making of wire. Originally, it was hammered out of gold, silver, or copper, and was used almost exclusively for the creation of jewelry or other fine decorative items for the wealthy.

The art of "drawing wire" was developed in Germany about 1350. As the development of metal quality improved, the art of wire drawing progressed as well.

"Iron wire" was first drawn in England in the late 1400's. Later, iron drawing processes were improved to the point where iron wire was readily available in Revolutionary times.

This availability allowed a suspension bridge to be built over the Schuylkill River near Philadelphia using iron wire wound into cables.

The uses of wire increased rapidly. In its relation to the subject of this book, the most notable use of wire was the invention of the Morse Patent of the electric telegraph in 1848. This patent sent an electrical signal over wire lines installed between transmitting and receiving stations.

By 1851, an estimated fifty companies in the United States and many more in Europe were using the telegraph in their business operations.

Because of these many miles of telegraph lines, it meant smooth wire was readily available at the time of the Civil War. It was inevitable that this smooth wire would eventually be used in war.

++++++

## CHAPTER 2. 1860-1865

As the Civil War escalated, both Union and Confederate forces had access to telegraph wire.

In fact, some of the success of the Union was attributed to a better communication system between command posts and the battlefront.

This improved system included several units of personnel trained in the installation and operation of the telegraph as well as battery wagons containing galvanic batteries to charge the wire for the signal.

The Union telegraph wire was unique for the time as it consisted of seven small wires wound together and insulated with a newly invented substance called "Gutta Percha," made of rubber.

This allowed the wire to be installed along the ground without diminishing its signal by grounding. The wire was carried on reels by both pack mules and wagons and could be unwound off the reel and rewound for further use.

The Confederacy also had access to telegraph wire, non-insulated, as railroads used the telegraph to schedule rail traffic. Telegraph lines also followed the Mississippi River to assist in the scheduling of the river boats.

Some wealthy plantation owners installed private telegraph lines to their homes to keep abreast of the cotton prices at the port markets.

So, wire was available for both sides of the conflict, and as defense fortifications were constructed, telegraph wire was utilized.

++++++

In 1860, South Carolina seceded from the Union of States. Major Robert Anderson commanded the First U.S. Artillery garrison at Ft. Moultrie on Sullivan's Island. Located in the mouth of Charleston Harbor, the area was made up of cottages and summer homes. Many were built close to the fort making adequate defense of the surrounding area difficult.

A palisade of pointed wooden stakes was installed along with other types of brush barrier to aid in slowing attack.

On December 25, 1860, Captain Abner Doubleday requested permission to purchase telegraph wire to entwine between the pointed stakes to further delay troop advance. His request was granted allowing him to purchase one mile, (5,280 ft.) of smooth wire.

However on the next day, December 26, concern for the garrison's troops resulted in their being transferred by boat to Fort Sumter located in the center of the harbor. The plan to entwine wire among the palisade stakes was never carried out.

In Captain Double day's personal account, he does not claim credit for the original idea of using wire in defense fortifications. This is the first recorded account of planned use of wire in warfare.

++++++

Accounts of entanglement wire use at Fort Sumter are vague but we assume with Captain Doubleday and Major Anderson being transferred to the fort, their original idea and the wire purchased was used in some manner.

The use of entanglement wire at Vicksburg can be pinpointed in time as detailed accounts are recorded. On May 15, 1864, Brevet Major General William Smith ordered his troops to take down a telegraph line running alongside the Turnpike Road. This was a private telegraph line running from the telegraph office in Vicksburg to a plantation nearby.

The troops removed the wire from in front of the Willis House near Dreary's Bluffs, and stretched it between the tree stumps and posts in front of their redan.

This barrier was successful as Union forces were twice repelled before succeeding in their final attack. Federal accounts stated, "Battle casualties were numerous with bodies piled upon the wire entanglements."

A painting entitled, "The May 19 Union Assault," by renowned artist There De Thulstrup, is now owned by The Seventh regiment Fund.

This artist's conception of the charge plainly shows telegraph wire trampled down at the bottoms of tree trunks and posts in the barrier.

These early attempts at incorporating wire into defense barriers merely complimented the use of other types of materials long used by the military. The effectiveness of wire use did not go unnoticed, but with the end of the war in sight, further accounts of use of wire in the Civil War have not surfaced.

++++++

Additional accounts of these first uses of wire in defense barriers includes the book, "Vicksburg, 47 Days Of Siege" by A.A. Hoehling. On page 128 he states, "General Shoup dictated to his official report - we are strengthening ourselves steadily and surely; are making in the ditch and in front of the redan, trough stockades with brush and wire entanglements."

Trough stockades are interpreted as shallow dry moats filled with anything at hand suitable to slow the enemy attack.

In Army Ordinance, Volume XV, No. 87, page 147, author Aiken Simons states, "Later, the Confederates used wire entanglements at Vicksburg."

The book, "Siege Of Vicksburg," by Richard Wheeler, page 165, states, "Some portions of our front were protected by abatis of fallen trees and entanglements of telegraph wire." Webster's dictionary describes abatis as, "a defensive obstacle formed of felled trees with sharpened branches facing the enemy."

From "War On The Mississippi," a volume from the Time-Life Series about the Civil War, a statement describes Union troops encountering "felled brush and *other* obstacles" in the assault of Stockade Reran.

Tour guides at the Vicksburg Battlefield Park stop the tour bus at the foot of Stockade Redan and recount, "In preparation for the coming Union assault, Confederate soldiers cleared the area of brush and timber, stacking it with the sharpened ends pointed to the enemy.

In between the remaining tree trunks, wooden posts were set firmly in the soil, and the ends sharpened with hatchets. Telegraph wire was entertwined between them."

A strange account of curious nature, is found quoted in The Encyclopedia Americana, International Edition, Volume 15, on page 147. It states, "Barbed Wire Land Mines and the Hand Grenade were innovations of the Civil War."

Since barbed wire was not invented until 1874, nine years after the end of the war, and the name "barbed wire" does not appear before that time, the statement is suspect to authenticity.

Perhaps the account was written later and the term was used inadvertently. The actual "barbed wire land mine" could have been an explosive device using cut pieces of smooth wire as shrapnel upon detonation. Whatever the explanation, this is the only reference to barbed wire in the Civil War accounts known to date.

++++++

In 1864, during the Civil War, iron making advanced dramatically. The introduction of The Bessemer Converter gave steel companies the power to control the quality of their product. Before the converter came into use, iron wire and other ferrous goods were subject to erratic composition, making the product subject to breakage. With this advance in steel making technology, the scene was set for the invention of barbed wire.

++++++

## CHAPTER 3. 1867-1875

Entanglement wire development halted with the end of the Civil War. However, progress on the development of fencing wire for the private sector continued.

A patent was issued to Lucein B. Smith of Kent, Ohio on June 25, 1867 for a wire to be hung between posts and containing points to thwart livestock passage.

An earlier patent issued to Alphonso Dabb on April 2, 1867 had barbs or points but was not suited for livestock control although it is the first U.S. Barbed Fence patent.

William D. Hunt, Michael Kelly, and Lyman P. Judson, all of the state of New York were issued patents for barbed wire before 1871. The designs were not successful because they were made by hand, very expensive, and were not well known to stockmen.

In 1873, Mr. Henry M. Rose of Waterman Station, Illinois made a barbed strip for his own use. This strip was not made of wire but was constructed of a one inch square wooden strip with sharp nails portruding outward every few inches. This rail hung inside a regular smooth wire fence and was successful in deterring livestock of all kinds.

Mr. Rose was issued Patent # 138,763 on May 13, 1873. He subsequently displayed his rail at the local DeKalb, Illinois County Fair. Three local citizens of the area viewed the display and each envisioned barbed wire as we know it today.

After returning home, Isaac L. Ellwood, a hardware merchant, Jacob Haish, lumber dealer, and Joseph Glidden, a farmer, all began working on their visions of barbed wire.

Glidden made the first application for a patent on October 27, 1873, but was not granted patent until November 24, 1874.

Haish was awarded three patents between January 20 and June 24, 1874. However, his application for patent came after the Glidden application date. In 1892, the U.S. Supreme Court ruled that Glidden owned the right of the barbed wire patent because of the early application date.

By the end of 1875, forty patents had been issued to inventors for barbed wire patents. Factories were numerous and the manufacturing process was constantly being mechanized.

Henry W. Putnam patented an automatic barbed wire making machine on February 27, 1876 and began producing large amounts of the new barbed wire.

Production records of barbed wire produced annually shows 10,000 pounds in 1874, 600,000 pounds in 1875, 2,840,000 pounds in 1876, and 12,860,000 pounds in 1877. Production grew by leaps and bounds after this date.

Also during this period, the U.S. Patent Office issued 8 patents in 1874, 19 patents in 1875, 39 in 1876, 37 in 1877, 32 in 1878, and 24 patents issued in 1879. This totals over 160 barbed wire patents of every conceivable design available for manufacture.

++++++

The success of the new barbed wire and the quantity produced made the invention available to everyone not only in the United States but on a world-wide basis. Within a few short years, stockmen all over the world were using barbed wire to some extent.

Although there were no major wars during this period, the world became familiar with the product and its potential, especially the military forces of foreign countries. With this availability, military strategists began to incorporate barbed wire into defense fortifications.

Research in this period has not revealed the use of wire barriers other than the regular livestock fences used by stockmen. Accounts of this use, if any, must be buried deep in military history as it has not been found to date.

++++++

## CHAPTER 4. 1876-1886

During the late 1870's and early 1880's, France began building fortifications to defend its borders against aggression. Among the many requirements demanded of its fortification engineers, was the need of a redoubt, (fortress), to resist explosive projectiles. Another asked that all entrances and exits be protected from attack. Many of these requirements were answered with the result being, war changed forever.

The book, Fortress- A History Of Military Defense, by Ivan V. Hogg, published by St. Martin's Press in 1975, recounts the following description of this French effort.

"In 1886, M. Mougin, a French Professor of Fortifications, devised a redoubt (fortress) meeting the requirements of the French military strategists.

Certain strategic sites, with unrestricted fields of fire, were chosen along the French borders. Heavy concrete and steel fortresses were constructed for the armament.

Soil was excavated around the fortress and piled over the concrete structures. An unscalable steel palisade fence with posts sunk in concrete was installed around the fortress.

The ditches or dry moats left from the excavations were filled with wire entanglements underlaid with land mines.

The approaches, (exits and entrances), were protected by removable wire devices and unlimited fields of fire from the new, recently developed machine guns."

Prior to 1886, the wire entanglements used by the French consisted of smooth wire entwined between wooden pickets about two to four feet above the ground. The depth and severity of these barriers depended on the terrain.

After 1886, barbed wire began to be used in new installations. Both smooth and barbed wire entanglements were sometimes wired to trigger land mines upon touch.

After 1890, installation instructions for barbed wire entanglements began to appear in various military manuals.

The manuals seemed to make efforts at justifying the use of vicious wire in war. One statement said, "We cannot be involved in the politics of war, the causes or reasons. We must concentrate on entanglement use."

Another statement about barbed wire defenses says, " It requires the attacker to engage and force his way into every stronghold no matter how small or insignificant." The use of barbed wire in warfare became a standard requirement.

As the turn of the century drew near, everything was in place to see a dramatic increase in the use of entanglement wire in warfare.

The new invention was readily available everywhere in the world, military strategists were aware of its capabilities, and the French and other countries had gained experience in constructing fortifications using barbed wire.

Most important of all circumstances, war loomed on the horizon. The next few years would see many confrontations between many countries. Each of these wars saw a dramatic increase of wire use in defense fortifications.

++++++

## CHAPTER 5. 1898-1900

The first of these confrontations started in 1898 with the Spanish American War. As most of this war was fought at sea, encounters with barbed wire were few.

There are some personal accounts but they do not describe wire barriers in detail. However, among the many illustrations and photographs of war scenes, the use of wire is portrayed.

The two volumes of Harper's Pictorial History Of The War With Spain, are considered by historians as the most accurate recording of this war. Several personal accounts contained within mention barbed wire.

One account states, "Barbed wire fences slowed troop movement." Another states, "Jungle growth entwined among barbed wire fences was a formidable barrier slowing progress."

The troops in Cuba were not equipped to cut through wire barriers. This fact was quickly corrected by the military by purchasing civilian wire cutting tools for issue to troops destined for battle in the Philippines. This is the first recorded instance where U.S. troops were issued wire cutting equipment prior to battle. Experience gained from troops encountering barbed wire barriers began to accumulate.

++++++

A study of the numerous illustrations contained in the Harper's volumes reveal eleven scenes showing barbed wire . Ten of these show standard livestock variety fences made of wooden posts and three or more barbed wires attached. All have jungle growth entwined.

The eleventh illustration is by the renowned artist Frederic Remington, who had been retained by Harper to portray scenes of the war. This scene clearly shows pre-planned entanglement barriers constructed around the fortress on top of San Juan Hill.

On July 1, 1898, at Santiago De Cuba, the Battle of San Juan Hill occurred. Remington's illustration shows six or more standard varieties of barbed wire fencing constructed in a parallel manner surrounding the outside walls of the fort.

Materials are wood posts set about six feet apart, with eight or more barbed wires attached in a horizontal configuration. Though crude and lacking imagination, the fences were successful in slowing American attack.

++++++

Accounts of the war in the Philippines also mention barbed wire fences of standard variety. No references to pre-planned entanglements are found to date.

Barbed wire use in entanglements during the Spanish American War was not significant. The fact it did slow troop advancement, and was described as a "formidable obstacle" by officers, added to the military experiences with barbed wire barriers. The issuance of wire cutters to the troops in the Philippines was the first known step in military counter-action.

++++++

## CHAPTER 6. 1899-1902

While the U.S. fought the Spanish American War, the British became embroiled in a South African war with the Boers.

The area of this action consisted of mostly farms and ranches, some of which used extensive barbed wire fencing.

The British military was behind in barbed wire entanglement use and experience. This was attributed to the fact that barbed wire use was prohibited by law in England on the grounds that it would lacerate livestock.

To date, the first known instance of wire use by the British was in 1884 when General Gordon entered Khartoom. He ordered smooth wire entanglements built in front of earth works alongside a river front.

Later, in 1888, a British Field Manual Of Field Engineering discusses and illustrates the construction of a barbed wire entanglement .

++++++

The South African War, or the Boer War as it is sometimes called, was the first war in which extensive pre-planned entanglement wire defenses were used.

Two books which recorded the details of this war are specific in describing the use of barbed wire barriers. To The Bitter End, by Emanoel Lee, is a photographic history of the Boer War. The Great Anglo-Boer War by Byron Falwell also tells the history of the war.

A personal account during the early part of the war states, "At Magersfontein, it appears that by chance, some barbed wire had been strung in front of parts of the Boer positions, but at this time it was not used deliberately."

This statement shows that the barbed wire encountered in the countryside was not placed to slow troop movement. As the war escalated, British military strategists devised plans to protect positions with wire barriers.

These plans called for the country to be divided into small areas by British fortified defense lines. This was supposed to prevent the Boers from joining together for mass attack.

The most important of these fortified lines was a railroad line running from the Cape.

Blockhouses, (forts) were to be built at selected intervals to guard the railroad and to prevent passage of the Boers to other areas.

In July, 1900, detachments of men were stationed along the railway to defend each section. A number of blockhouses were erected near bridges. Trenches were dug and reinforced with stone walls, while barbed wire entanglements were built in front of these walls.

To complete these defenses, additional barbed wire barriers were placed between and adjacent to the blockhouses. An excellent description of these entanglements is contained in the following account.

"These were not simple fences but a mass of protective obstructions fashioned into carefully worked out, complicated and seemingly impregnable defensive barriers. The fences were further strengthened by twisting together barbed wires and using strong annealed wire which could not be severed with simple wire-cutters. The strands were held together by thick wire and metal stays which extended down into the ground where they were anchored with sand bags. Bells were attached to the fences to sound warning if the wire was touched at night. Stationary loaded rifles covering predetermined segments of the veld were fired automatically via trip-wires."

Other observations from personal accounts mention that prisoners were held behind barbed wire stockades. Photographs show families of prisoners were held in open camps with a single barbed wire around the parameter. No one was permitted to go beyond the wire.

Much of the barbed wire design shown in photographs appears to be a galvanized ribbon with barbs. This Brinkerhoff ribbon wire was popular with the farmers and ranchers of the area.

Experience in using entanglement wire for military defense grew by leaps and bounds as the strategists planned for the future. Each wire encounter seemed to increase the respect the military had for barbed wire.

++++++

## CHAPTER 7. 1902-1905

The book Fortress, by Hogg states, "The notable thing about defenses such as barbed wire is that it requires the attacker to engage and force its way into every stronghold no matter how small or insignificant."

Russian military strategists evidently took this quote to heart as they began constructing massive defense works in Tsintao in 1902. These defensive positions included Kwantung Peninsula and Port Authur.

Two books, The Japanese-Russia War, by Sydney Tyler, and Tide At Sunrise, by Denis and Peggy Warner give numerous accounts of the use of barbed wire entanglements in this war.

A study of these accounts give a vivid picture of the size and scope of the barriers. An early 1902 statement says, "In front of the defenses, a scarped ditch was dug and the bottom strewn with barbed wire. Landmines were laid with electric fences wired to set off the mines when contacted."

Various personal accounts state, "A network of barbed wire fronted the assailant as he rushed to attack." Another added, "In front of my position, the barbed wire entanglements stretched across the isthmus." An officer explained, "The Eastern approaches consisted of two fences of barbed wire encircling a mine field."

Other barriers were used in combination with wire as this account states. "Rifle pits and trenches were dug with all approaches guarded by barbed wire entanglements." Another account says, "The triple lines of trenches were protected by ten foot barbed wire entanglements over pits with sharpened wooden spikes at the bottom."

The extent of the defenses was described as, "The approaches were rendered almost inaccessible by the placement of a triple row of barbed wire entanglements."

As could be expected, the extensive use of many types of barriers as well as barbed wire had a conclusive effect on the battle. Casualties by the attackers were enormous.

A war reporter wrote, "The forward troops of the attackers reached the wire entanglements and began hacking at the wire with swords. The rear elements soon caught up. When the whole lot became involved with the wire, the Russians opened fire with their Maxim and Gorloff machine guns. The result was massacre."

A Russian soldier stated, "The enemy troops were contained at the edge of the wire. Each time they returned to the wire they were cut to pieces by gunfire."

Another correspondent added, "Rush after rush was made by the Japanese but were hurled back before they could lay their hands on the wire. The entanglements proved to be a deadly obstacle to the enemy."

It seemed the advantage of the barbed barriers would carry the day for the defenders. But the aggressors were many and as the battle wore on, tactics and tools were developed to combat the wire.

The first known wire related title, "Pioneer," was originated as soldiers volunteered to crawl forward to cut the wire. Equipped with wire cutters and crude iron shields which were pushed ahead as they crawled, these men were the first of specially trained detachments used to combat the entanglements.

Other means used wooden poles and lumber as well as sheet iron laid across the top of the wire to mash it down for troops to progress.

In the end, as the Japanese finally won out, much of their advancement through the wire barriers had been over the bodies of their comrades who had fallen crossing the wire.

This deadly example of the effectiveness of properly placed barbed wire entanglements echoed through the minds of military strategists throughout the world. Defensive planning had now entered a new phase.

++++++

## CHAPTER 8. 1905-1913

With the end of the Russo-Japanese conflict, the world was temporarily at peace. But for some countries in Europe, it was an uneasy type of peace. In expectation of threats to come, defense work along their boundaries quietly continued.

The terrible effectiveness of entanglement wire against the Japanese at Port Authur caused a flurry of obstacle wire development in many major countries. Engineers were busy designing and testing all types of ideas both for installing and combating wire barriers .

Some of the major armies of the world published manuals and specifications by which to manufacture and install entanglements quickly.

Specifications governing the designs and materials of warwire were issued as well as requirements for packaging and shipment.

The wire specifications included size, gauge, strength and barb design, as well as how the wire would be tested. Packaging requirements included ease of shipping, handling and damage prevention, and ease of use by the troops in the battle areas. Transportation costs were always a factor.

Once these specifications were accepted, elaborate contracts were drawn, bid requirements published and actual bids for manufacture of the products were solicited from the steel companies. The various military branches began to stockpile entanglement related materials.

One of the earliest editions of Warwire specifications on file in The U.S. Army Military History Institute, is the 1917 pamphlet No. 460, entitled Special Specifications For Steel Wire.

It is not the first edition as it refers to an earlier Ordanance Pamphlet No. 445. In addition, a revision schedule printed on the front page shows that specifications for steel wire for military use were updated and revised almost every year since January 14, 1904. This date could possibly be the first time for specifications.

The earliest Field Engineering Manual on file at the Institute is an April, 1918 pamphlet No. 792, entitled Wire Entanglements Addenda No. 1. It is a reprint of a pamphlet prepared by The General Staff Headquarters, American Expeditionary Forces, France, February, 1918. There is also reference to an earlier National Training Pamphlet No. 150. Until further documents are found, it is believed that January, 1904, is the first time an entanglement field engineering manual was published by the U.S. military.

These publications tell us that although the world was basically at peace, entanglement wire development was ongoing. During the period from 1905 to 1918, wire specifications and design were revised almost annually. Invention, testing, and manufacture of entanglement wire related devices were being stockpiled. Also, engineering field manuals were published and the military of many countries were training soldiers to install and combat entanglement barriers.

++++++

## CHAPTER 9. 1912-1918

As barbed wire entered into the new role of a much used war weapon, it is ironic it remains an instrument of defense only. Never has barbed wire been armored for offense.

In the book, The Wire That Fenced The West, by Henry and Frances McCallum, they state, "The part which barbed wire played in two world conflicts was of greater significance than is usually supposed."

The McCallums reinforce this statement with an excellent description of what happened in the first half of World War I. They write, "Miles of wire were strung, bales of it scattered, rolls of it were tangled across the battlefields of Europe. Ground troops found it impossible to make their way through the matted web of wire that was so sharply barbed soldiers could not take hold of it to cut it. It was said that barbed wire was the most formidable obstacle a modern infantry had to contend with."

Modern day historians cannot explain why so little was written in the books of WWI about the use of entanglement wire. Page after page attend minute details of the war but seldom mention barbed wire. Yet, right along with these details, accompanying photographs show extensive wire entanglements of every imaginable description. It's almost as if a moratorium had been declared against mentioning barbed wire.

Another impression perceived when doing research is that historians and correspondents were somehow trying to preserve the old ways of war and ignore the new ideas and materials.

A comparison of officer's accounts and the personal accounts of the foot soldier on the front lines, suggest the officers were fighting a different battle than the foot soldiers.

A quote by Marshal Foch in 1926, might answer this question. He said, "The military mind always imagines the next war will be fought on the same lines as the last war. This has never been the case, nor will it ever be."

++++++

At this particular time in history, especially in Europe, most officers were of the old school. Some countries still allowed commissions to be purchased and actual war experience did not seem to matter. Those who came up through the ranks fought strictly by the book no matter the consequences.

World War I was such that thousands of lives were lost by bad decisions made by these unseasoned officers. Failure of these leaders to adjust to new innovations of war was devastating to the foot soldier. It was a well-known fact that in the past, wars were won by whoever had the most men to expend. The development of machine guns and barbed wire became the equalizer between this imbalance.

Most of the older military leaders of Europe did not study the lessons of Port Authur, and were not aware of the effectiveness of barbed wire entanglements and trenches. This ignorance costs thousands of lives needlessly.

With the exception of Germany, each of the major countries involved in this Great War believed they were fighting a war of defense. Hence, in order to *defend* themselves, they attacked each other.

All of these shortcomings added up to a deadly result. The expected short war extended for years, tactical blunders caused the loss of thousands of lives, tremendous expense, and the devastation of their lands.

++++++

WWI, sometimes called the Great War, was popular in the beginning because of the nationalism of the many countries involved. Also, it was predicted the war would be of a short duration, only a mild confrontation needed to straighten out some boundary disputes. Time would prove these thoughts to be wrong.

The summer of 1914 proved to be a busy time in the war business. On July 28th, Austria declared war on Serbia. On August 1st, Germany declared war on Russia, and on August 3rd, declared war on France. On August 4th, Great Britain declared war on Germany. At this time, all major countries of Europe except Italy were at war.

However, as these supposedly short conflicts wore on without swift conquest, other countries and much larger territories became involved. By 1915, all of Europe was engaged, and of the great powers outside Europe, only the U.S. was not implicated.

The Great War continued to grow and live up to its name, until it involved every major state on the globe. At the final peace conference in 1919, thirty-two "victorious nations" were represented, and twenty-two were non-European.

Because of the many countries involved in WWI, the use of entanglement wire and the lessons of Port Authur became well-known the world over. This somewhat new innovation now became a standard military tool of war, and all countries used it to their best advantage.

++++++

Both Great Britain and Germany were world leaders in steel and iron production. This fact provided their own armies and the armies of their allies all the barbed wire and related devices needed for fortifications.

European military strategists took the lessons of Port Authur to heart as row after row of trenches and entanglement wire enclosed vital fortifications. Added to these barriers were the new machine guns covering every square foot of the approaches leading to the wire barriers.

Each charge of the aggressors seemed to end in massacre, leaving their comrades hanging on the wire. Heretofore, unknown morale and chaos problems arose at the fronts. It was at this time the age-old war tactic of frontal attack was rendered obsolete.

With the troops refusing to participate in a frontal attack, they were left in "trench deadlock" unable to pass beyond the wire entanglements to reach the enemy.

One officer's analysis of this situation stated, "Machine guns and barbed wire have made infantry advance hopeless and use of the Cavalry futile."

A bewildered veteran penned in trenches near the battlefront described the dilemma, "We fall in death without ever having seen the enemy. War has become man fighting against materials, and we fight with the sensation of striking out at empty air."

The years of 1915 and early 1916 saw the heaviest fighting among men, and the most concentrated output of war materials known to date. The use of entanglement wire and machine guns had created a definite advantage for defensive warfare.

++++++

However, all things change, especially war. The pendulum always swings back. In 1916 many innovations were introduced to offset the advantages enjoyed by the defense. This breathed new life into offensive forces and the war continued onward.

Improved armament in artillery, machine guns, and explosives came onto the war scene. An unending supply of ammunition and fresh troops created an entirely new type of warfare.

Credit for the start of this turnaround goes to the German military who devised a maneuver that proved to be extremely effective against barbed wire entanglements and trenches.

This maneuver was simple. German military strategists pinpointed an area in the enemy front lines, laid out the parameters, and called the troops back from this area. While the troops rested and refitted, heavy armament was concentrated on the front line area.

Grid patterns of targets were laid out, followed by hours of bombardment saturating the area with explosive shelling.

Just before the shelling ended, the fresh troops were in position to attack. When the bombardment stopped, the troops rushed into the area.

This tactic was successful as the trenches were reduced to mere shell holes, and the barbed wire entanglements were piles of tangled wire. Enemy survivors were either dead or disoriented from the blasts.

The aggressors forged ahead until they began to meet resistance. There they stopped, rested, and called up the armament for shelling.

Other innovations to combat barbed wire began to appear. It is unclear who actually created some of these ideas, but Germany is credited with using mustard gas and flame throwers.

Smoke screens, used in combination with shelling and troop movement, allowed passage through the entanglements.

Machine guns and other rapid-fire weapons became more mobile and easier for the offense to move into position for use.

September 15, 1916 became a red letter day in combating entanglement wire barriers. The British sent fifty tanks into battle.

These armoured vehicles were a new idea, not fully tested, and susceptible to breakdowns. But the British were desperate for anything to boost troop morale and achieve a victory. The new tanks were successful in this endeavor.

The British sent out fifty tanks from a gathering point behind the lines. Only thirty-two reached the front lines as the others suffered breakdowns of one sort or another. As the battle surged forward, twenty-four tanks were moving. Others continued to be disabled or bogged down.

Only four of the original fifty tanks finished the victorious battle and entered Flers with cheering troops marching behind.

In spite of the small number of tanks to survive, they had proved their worth against wire and trenches. Warfare entered a new phase.

The final incident marking the equalization between defense and offense, came when the most powerful fortress of the time fell to the Germans.

Located at Verdun, this magnificent defense fortification contained concentric rings of over sixty mighty underground fortresses.

It was protected outwardly by an intricate network of trenches, and wave after wave of wire entanglements throughout its length.

The German tactic of bombardment, followed by troop assault, combined with flame throwers to clear the enemy from underground fortifications was successful. The fortress fell quickly, and like the new tanks, these events ushered in a new phase of warfare.

From humble beginnings at Vicksburg using salvaged telegraph wire, entanglement wire use had grown to the point that it deadlocked troop advance on the battlefront.

Now, new tactics and new innovations had penetrated the barriers. Wire entanglements were still a valuable tool for defense and would be used extensively in the future. However, its effectiveness had been diminished.

The record of accomplishment of barbed wire used in warfare is significant. Although used only as a defensive device, its effectiveness had changed warfare.

For example, the age-old tactic of frontal attack by the infantry and cavalry was rendered obsolete as troops waited for the artillery to act.

The role of artillery was expanded into the major instrument of attack. Shelling was more effective as the explosives gained power and range. Targets could be pinpointed with such accuracy that saturation bombardment became a science. All of this came about because of the use of entanglement wire and trenches.

The machine guns used so effectively by the defense in combination with barbed wire, became more mobile as the offense carried them into battle.

Though warfare still needed good leaders and well-trained troops, there was a growing focus of materials fighting materials.

The end of WWI finally brought peace to the world. As the many countries involved rested and licked their wounds, the military strategists studied the lessons of this war just as they had studied the lessons of the Russo-Japanese War.

Entanglements could no longer hold back the aggressors on the front lines, however, it could still designate boundaries, create safe havens behind lines, protect supply depots from sabotage, and confine prisoners of war safely.

With these thoughts in mind, the peacetime military strategists went to work designing the new role for entanglement wire use in warfare.

++++++

## CHAPTER 10. 1919-1936

The end of the Great War was a welcome relief to all involved. The "short duration" expected in the beginning, turned into years of savage fighting and destruction.

The cost in lives, money, and resources was staggering. The countries where the battles actually took place were devastated and stood on the brink of economic disaster.

Now all sat back, licked their wounds, and contemplated the results of the war. Sadly, after all was said and done, some of the original reasons for the war were still unresolved.

By the late 1920's, some countries quietly began adding to fortifications along their borders. The military strategists of all countries studied the lessons of WWI, and began development of the war equipment and tactics to be used in the next war wherever and whenever it should occur.

The successful innovations, both in tactics and weapons, coming out of the second half of the war were numerous and significant.

Most notable among new tactics was the expanded role of bombardment before attack. The improved range, velocity, power, and mobility of artillery reduced barbed wire entanglements to mere piles of wire and trenches to shell holes. Enemy troops and other fortifications were left in a state of easy capture.

Now, added to the bombardment tactic, was the armoured tank introduced by the British, and the airplane both of which showed great promise for use in warfare.

It became generally accepted that the old immovable fortifications, wire entanglements, and trenches would play a lessor role in defense.

"Rapid Deployment" became the new term describing tactics and equipment for the future. With fast moving armoured vehicles, followed by trucks hauling fuel and supplies, the next war would move at a fast pace.

Because this rapid deployment warfare moved so fast, ground troops faced the problem of keeping up. The new equipment could easily out-distance, isolate, encircle and entrap lagging ground troops, resulting in an increase in numbers of prisoners of war.

There had always been prisoners captured during war but recent treaties adopted in Geneva, now required new responsibilities of the captors.

This anticipated increase in prisoner numbers, plus the new responsibilities, would have an impact on the use of barbed wire. To place this prisoner of war problem in proper prospective, a review of the development of "The Laws Of War" is neccessary.

Numerous sections of this long, tedious code of ethics for war, pertain to the humane treatment of prisoners.

++++++

A quote from the book, The Laws of War, by Reisman and Antoniou, describes war times before the turn of the century. "The unofficial rules of war were respected and enforced in an ongoing process of reciprocation and retaliation."

Simply put, this says countries that committed violations would excite retaliation in such a manner that respect for the norm was re-established.

By the turn of the century, a world- wide peace movement sought to condemn war or temper its severity when it occurred.

At times, Arms Control Agreements were sought between countries to maintain power balances. Because of the tremendous cost of modern warfare, lower levels of military parity were sometimes mutually agreed upon.

Treaties, which are formal written agreements between two or more countries, are a relatively new and popular development of war.

Through the efforts of The International Committee of the Red Cross, founded in 1894 in Geneva, numerous arms agreements and treaties have been adopted by most countries of the world.

Starting with the "Hague Convention" in 1907, other conventions followed down through the years until modern times. These cumulative documents are known as The Geneva Accords, The Geneva Laws, or The Laws of War.

Among the many aspects addressed in the accords, the humane treatment of prisoners of war was continually updated. This raised the status of war prisoners to the point where captors were required to protect and provide adequate care for them.

This is quite a change from the past when prisoners were considered a dangerous burden to captors and were sometimes put to death.

Later, prisoners became more important as the countries at war exchanged prisoners. Added value came about as captors used prisoners as slaves to further their own war effort.

As warfare became more devastating, participants realized that returned prisoners became the seeds of the future.

As the lot and worth of prisoners of war improved, the responsibilities of the captors increased. Questions of how to hold, confine, feed, and house prisoners arose. Most of the answer was simple, use barbed wire.

It was cheap, plentiful, quickly installed, and effective as a prisoner deterrent. Because of these reasons, barbed wire stockades and temporary detention centers made from barbed wire became standard tools of war.

++++++

## CHAPTER 11. 1914-1946

The fate of prisoners of war had always hinged on the battlefield circumstances of the captor troops at time of capture. If time and resources were available with which to tend the prisoners, it was done. If not, often the prisoners were executed or abandoned to facilitate further aggression.

These massacres and abuses were known only to the participating troops and bystanders. Because it was considered an act of war, most overlooked the incidents. The poor communications of the time kept the facts of these horrors on a local basis.

At the time of WWI, communications had improved to where little of history could be hidden away. As this war escalated, atrocities committed to captured enemy troops began to surface.

The worst incidents happened on the German/Russian front where fast paced German forces entrapped thousands of Russian soldiers.

German forces were not prepared to tend the needs of these prisoners. As a hard winter season set in, most prisoners died of starvation and neglect. This pattern happened on other fronts, but to a lessor degree.

As the facts of these horrors became public knowledge, world-wide humanitarian efforts were started to improve the lot of POW's.

These efforts, spearheaded by the Red Cross, resulted in most all countries of the world signing treaties and agreements expanding the humanitarian aspects of prisoners of war. Only Asian countries such as Japan refused to sign.

These cumulative treaties, known as The Geneva Law, are being continually added to or upgraded as needed by the demands of war.

As these agreements became international law, the fate of POW's improved dramatically. Food, shelter, and medical treatment were given first priority when possible. The method of confinement was not addressed, however, the method most used by all countries was stockades of wire.

Stockades could be quickly and economically built at any location. Captured prisoners could be used to build such facilities. Little was lost if the urgencies of war required its abandonment. Barbed wire made excellent temporary holding facilities for prisoners.

Stockade construction varied in design. Open areas used ten foot high posts with barbed wire stretched horizonally a few inches apart and attached with nails or staples.

If existing buildings or walls were incorporated, barbed wire was used around the top to discourage climbing over. Movable barbed wire barricades blocked any openings.

If plaza type areas were utilized, street and alley exits were also blocked with temporary barriers made of barbed wire.

++++++

During the peacetime period between WWI and WWII, efforts to improve POW conditions continued.

A basic legal procedure was devised to facilitate exchanges of prisoners between warring countries.

Another procedure presented guidelines for working prisoners to aid in the captor's war effort.

An aid program, to be supervised by the International Red Cross, permitted Care Packages to be delivered to needy prisoners.

Correspondence written by prisoners to their home country was allowed with only classified information censored.

Facts about prisoner abuse and ill treatment could not be censored from prisoner's letters. This alone prevented abuse to prisoners as the captors feared retaliation against their own comrades held in the enemies hands.

With these and many more agreements now international law, future prisoners taken in war were assured of better treatment than in the past wars.

++++++

Though all countries that signed the Geneva Accords were fully aware of the new laws concerning prisoners of war, none were prepared for the dramatic increase in POW numbers during the early part of WWII.

The rapid pace of modern war machines left thousands of foot troops vulnerable to entrapment and capture. This was especially true in North Africa where mild terrain allowed vehicles to progress at top speed.

By early 1942, Britain accumulated over 50,000 prisoners with all facilities filled to capacity. The U.S. agreed to take these prisoners to alleviate overcrowding.

In November of 1942, when the U.S. entered the North African campaign, military officials were told to expect maybe 10,000 POW's.

By early 1944, over 100,000 POW's had been shipped to the U.S. At the end of the war in 1945, U.S. POW camps held over 425,000 men.

How were these prisoners held? Mostly behind barbed wire in some configuration or another. Hard-core Nazis and other high security risk prisoners were held in military prisons behind high walls topped with barbed wire.

Medium security stockades called Main Camps, totaling 155, were chain-link enclosures reinforced with various barbed wire designs.

Minimum security installations, called Branch Camps, totaled over 500 by the end of the war. Security measures were extremely varied.

All were temporary work camps and utilized whatever facilities made available. Some of the remote farm camps had one barbed wire around the parameter and few guards.

The POW's lot had improved to the point that few wanted to escape. The only exceptions were the fanatical Nazis.

++++++

Ironically, massive bombardment and dynamic mobility of war machines rendered barbed wire entanglements obsolete. Yet, these same forces created a dramatic increase in the European Front prisoners of war.

To properly tend the needs of these prisoners as required by law, massive construction projects were needed, all requiring barbed wire as security. Although barbed wire was eliminated in one use, another need arose.

The war in the Pacific was a different matter entirely. Japan did not sign the Geneva Accords and did not recognize the International Laws of War.

Most allied prisoners taken by the Japanese suffered death either in crude jungle prisons or on death marches.

The fanatical Japanese soldiers chose to fight to the death so few prisoners were taken by the Allies.

There was little entanglement wire used in the Pacific war. Barbed wire use was limited to barriers around supply depots, rest areas, and temporary military installations.

As the need for this type of barrier grew, experimental metal tapes were tried. These had razor points and were contained in canisters. This period also introduced pre-formed concertina coils of entanglement wire.

++++++

## CHAPTER 12. 1950 - 1953

The Korean War, often called the "forgotten war," heralded the new term "limited war" in which democracy tries to limit the spread of communism in the far corners of the world.

The larger aspects of this war have been recorded, although historians believe a definitive study is wanting. Our research in this period proved frustrating in comparison to other periods of war studied.

A careful study of the day-to-day operations of the U.S. Army Infantry which furnished 86% of the U.S. troops in the war, revealed no mention of barbed wire entanglements, stockades, or razor wire.

A careful study of the details contained in photographs of these operations also revealed no signs of barbed wire use.

If barbed wire or any of its related devices were used in the Korean War, it was on a very limited basis.

++++++

## CHAPTER 13. 1964 - 1973

The Vietnam War was another "limited war" waged to limit the influence of communism in a foreign country. It eventually became the longest undeclared war in U.S. history.

Research efforts of this period of war were opposite those of the Korean War. It became an effort to glean through reams of information selecting certain items pertinent to barbed wire entanglement use .

An excellent article published in the October, 1967 issue of The Barbed Wire Times, touts the use of barbed wire in South Vietnam. The following excerpts are quoted from this article.

"On September 7, 1967, Secretary of Defense Robert S. McNamara ordered a barbed wire fence strung across the northern border of South Vietnam to limit the constant flow of Red infiltrators and supplies across the demilitarized zone.

Military experts state this fence will be barbed wire of the Military Concertina type and will also employ the use of an electrical barrier is to serve primarily as an early warning device rather than a solid barrier. It is hoped the fence will make crossing the border tougher."

Further information provided proof of the type and variety of the wire used in this fence. A dust and mud covered sample was sent home by Master Sergeant Phillip A. Knonenberg, U.S. Marines 3rd Division. The eighteen inch piece of wire was a 4 point Burnell variation known to servicemen as Concertina Wire.

Military installation photos accompanying the article shows ten to twelve feet high parameter fences with concertina coils at the top and backs of the structures. Varied widths of apron entanglements made from standard barbed wire were made in front of these main fences. The photos were taken at Phu Bui, Vietnam.

Another follow-up article in the May,1968 issue of The Barbed Wire Times states, "The Pentagon refuses to reveal the purpose of its recent contract award to Republic and Kaiser Steel corporations. These contracts are for materials necessary to build the McNamara Line between North and South Vietnam.

Republic Steel of Chicago is being paid $1,914,000 for 200,000 spools or 50,000 miles of barbed wire, enough to string 1,000 strands across the 40 mile border.

Kaiser Steel is being paid $3,348,415 for more than 5 million fence posts to support the wire."

These statistics on the amount of materials needed for entanglement fencing reinforces the importance of barbed wire used in warfare.

It is odd that barbed wire use in Korea was negligible while surging to major proportions in the later Vietnam War.

++++++

Razor wire had been in the experimental stages for several years. Military strategists estimated the effectiveness of razor wire was five times greater than conventional barbed wire. However, handling and installation problems plagued its use.

These problems were overcome by developing a packaging cannister that held the wire coils safely for shipment and installation.

At the installation site, the cannister was handled like an automobile tire, weighed 40 lbs, and acted as a carrier/handle for unwinding.

To lay out and construct a barrier with the razor wire was easy. A soldier drove a stake or set a post, anchored the wire ends to it, then lifted the cannister and walked away.

The metal tape, (razor wire) fed outward in two separate lines coiling instantly into concentric spirals. This one cannister of razor wire created an instant barrier 76 feet long and 2 ½ feet high. The spiral coils could be doubled in a parallel manner or stacked for additional width and height.

The article speculated the new razor wire would be sent to Vietnam immediately to combat and impede enemy troop movement.

Barbed wire use was liberal during the Vietnam War because of the guerrilla nature of aggression. The Viet Cong made persistant attempts to infiltrate base areas for sabotage and personal attack. The increased effectiveness of razor wire saved lives and supplies.

++++++

A check of Pentagon contract awards shows that in addition to the materials listed as needed to complete the McNamara Line, the defense agency bought 1.3 million coils of concertina wire in 1967 at a cost of $14.5 million.

In addition to the concertina wire, the agency purchased over 340,000 spools of standard entanglement barbed wire for aprons and other barriers. The contract cost was $3.2 million.

They issued contracts for over 15 million steel posts at a cost of $13 million.

Historians agree that entanglement wire use diminishes when battle front terrain allows rapid deployment tactics. But experience has proved when this tactic is used, POW numbers increase drastically requiring more temporary and permanent barbed wire stockades.

If the battlefront terrain is jungle or swamp preventing rapid deployment tactics, entanglement wire use increases as the foot soldier becomes the major tool of aggression.

It has also been proved that entanglement wire use, both standard barbed wire and razor, increases when the nature of warfare consists of guerrilla excursions by the enemy.

Interestingly, with all the improvements in armament, weaponry, communications, tactics, and technology, the age-old tool of barbed wire is still effective in defense and confinement.

++++++

## CHAPTER 14. 1973 - 1998

In each of the wars since Vietnam, the action was so fast-paced that barbed wire was not needed until after occupation. Parameter fencing, checkpoint reinforcement, and temporary stockades required limited use of barbed wire.

The Persian Gulf War was conducted on mild, desert terrain allowing rapid deployment tactics supported by aircraft.

The photos of this war show the use of razor wire in temporary stockades holding thousands of prisoners of war.

++++++

The facts and experiences presented in this history text lead us to believe that as long as there are military positions to be protected, and as long as prisoners of war are captured, there will be a need for barbed wire in warfare.

Late information found;

As "Warwire" was undergoing final editing, this bit of information was found. A TV documentary of the life of Pancho Villa told of his use of barbed wire against the Mexican Government's Cavalry forces in 1916.

We quote the written account, "In order to counter the government's cavalry forces, Villa strewed the lines in front of his own troops with barbed wire. When the on-rushing government cavalrymen's horses hit the barbed wire, they were stopped in their tracks, often dislodging the rider, or freezing in place, to avoid further injury to their legs. Villa's forces then proceeded to cut the troops down with the new type of machine guns created for use in WWI."

Villa was known for his rebel cavalry which had given him his greatest victories against the government. His cavalry was said to be the greatest and most famous in the revolutionary armies.

Ironically, his use of barbed wire against the government forces, contributed to the demise of the cavalry as an effective fighting force in modern warfare.

## SUGGESTIONS FOR FURTHER STUDY

### BOOKS


| | |
|---|---|
| Vicksburg, 47 Days Of Siege | By A.A. Hoehling |
| Army Ordinance Vol. XV No. 87 | By Aiken Simons |
| Siege Of Vicksburg | By Richard Wheeler |
| War On The Mississippi | By Time-Life Series |
| Barriers | By Campbell & Allison |
| Fortress, A History Of Military Defense | By Ivan A. Hogg |
| Harper's Pictoral History Of The War With Spain | (two volumes) |
| To The Bitter End | By Emanoel Lee |
| The Great Anglo-Boer War | By Byron Falwell |
| The Japanese-Russian War | By Sydney Tyler |
| Wire Entanglements Addenda No. 1 | U.S. Army |
| The Wire That Fenced The West | By Henry & Francis McCallum |
| The Laws Of War | By Reisman & Antoniou |
| Nazi Prisoners Of War In America | By Arnold Krammer |
| The McLean POW Camp | By Delbert Trew |

THE U.S. PATENT OFFICE

THE U.S. ARMY MILITARY HISTORY INSTITUTE

# SECTION I.

## WARWIRE DESIGNS

**Names and illustrated drawings of barbed wire known to have been used in warfare.**

# INTRODUCTION TO SECTION 1.

**This section is devoted to illustrations and patent information about barbed wire design. Minute details of each wire may not interest the average historian, however, to the wire collector these details are critical to correct wire identification.**

**Listing patent numbers and other data is the only way to authenticate origins of wire. In order to understand the idea and wire design of the inventor it is sometimes neccessary to study the actual patent application.**

**Some of the wires shown were never manufactured. Others were only made in small quantities. Just because a wire was invented and patented, did not mean the military would use it. What some wires lack in common sense design, they make up in uniqueness and were issued a patent.**

**From the first use of smooth wire in the Civil War, until the turn of the century, obstacle wire was limited to the standard livestock varieties. Some were twisted together in combination to improve their viciousness and strength. Heavy wire or metal stays were added to hold them upright.**

**Remember in many instances, entanglement design was the act of desperate men using anything at hand to create a barrier to approaching enemy. This type of barrier could contain any combination of wires or other such materials that lay at hand.**

**After 1900, military strategists began to specify the materials and designs needed for obstacle wire. Obviously these specifications included more barbs, of a sharper, longer, more vicious design, and strung on lines of stronger tempered steel.**

**The next development consisted of pre-constructed coils of wire for installation at the battle site. This speeded up construction and cut down on the time installing troops were exposed to enemy fire.**

**The last big change in the use of Warwire is the adoption of "razor type designs." This wire is lighter in weight yet stronger and more vicious.**

**Readers may note that many of the wires refer to the same patent number. Most of the pre-razor wires used by the military, were made as variations of the original livestock design patents.**

WWI ENGLISH RIBBON WIRE

A flat ribbon, four point, riveted wire, patent # and patentee unknown.

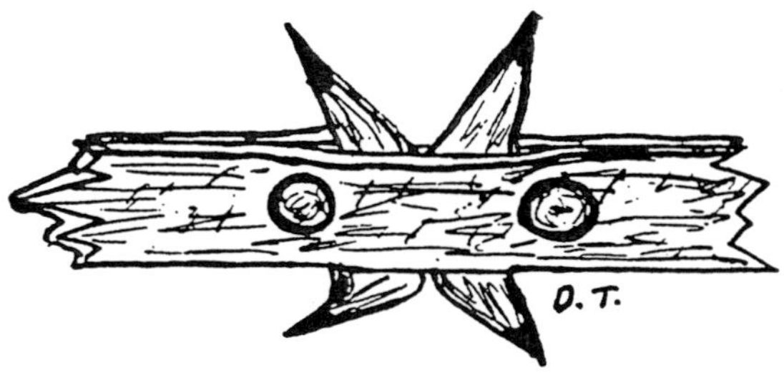


PAT.#6914(variation)

Patented Feb. 8, 1876 by Joseph Glidden. A single round, rippled strand with long four point barbs. May be clipped into double or triple parallel lines for use.

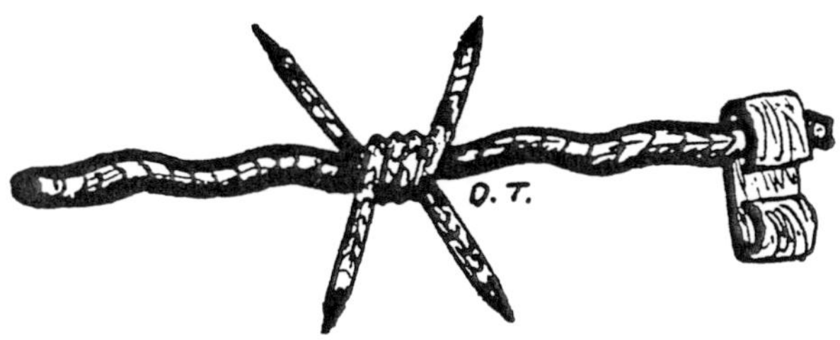


PAT.#191,001(variation)

Patented May 22, 1877 by Aaron Nellis. A flat metal or wood strip with vicious spring steel clip barbs. Known as U.S. Anti-personnel wire.

**PAT.#6914(reissue-variation)**

**Patented Feb.8,1876 by Joseph Glidden. Shown here in a parallel version used in making Concertina coils ready for installation.**

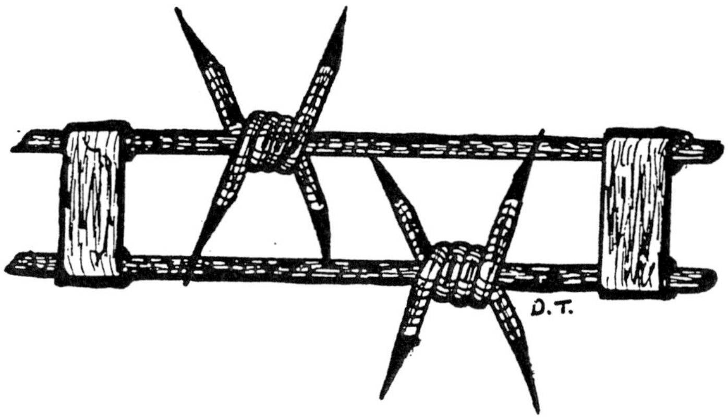


**PAT.#214,095**

**Patented April 8, 1879 by J. Brinkerhoff. Flat metal, galvanized strips, can be twisted, showing a variety of barb designs.**

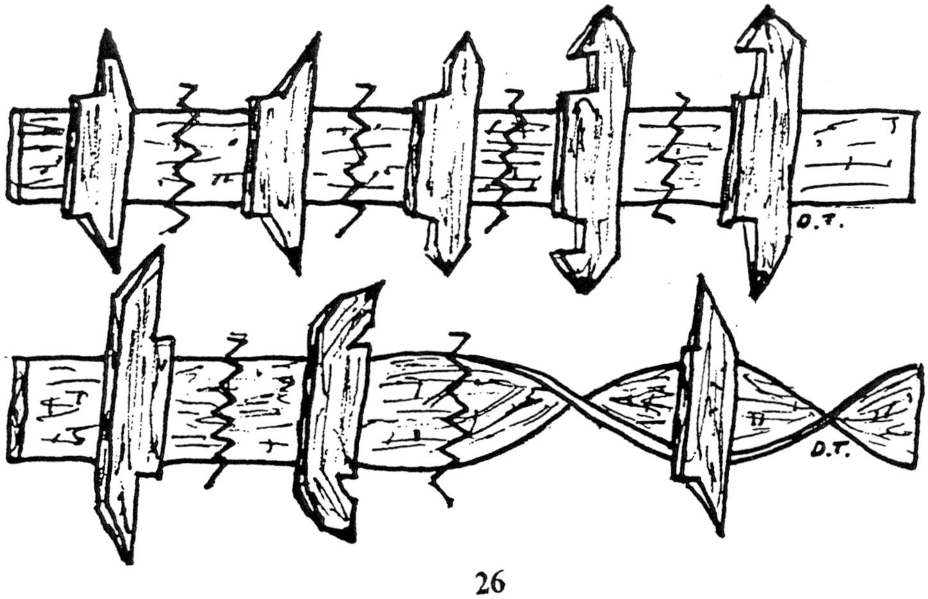

PAT.#1,281,697

Patented Oct.15, 1918 by Thomas Sturtevant. Fishhook barbs attached at factory proved useless as wire could not be unrolled. Field attached hooks were then offered.

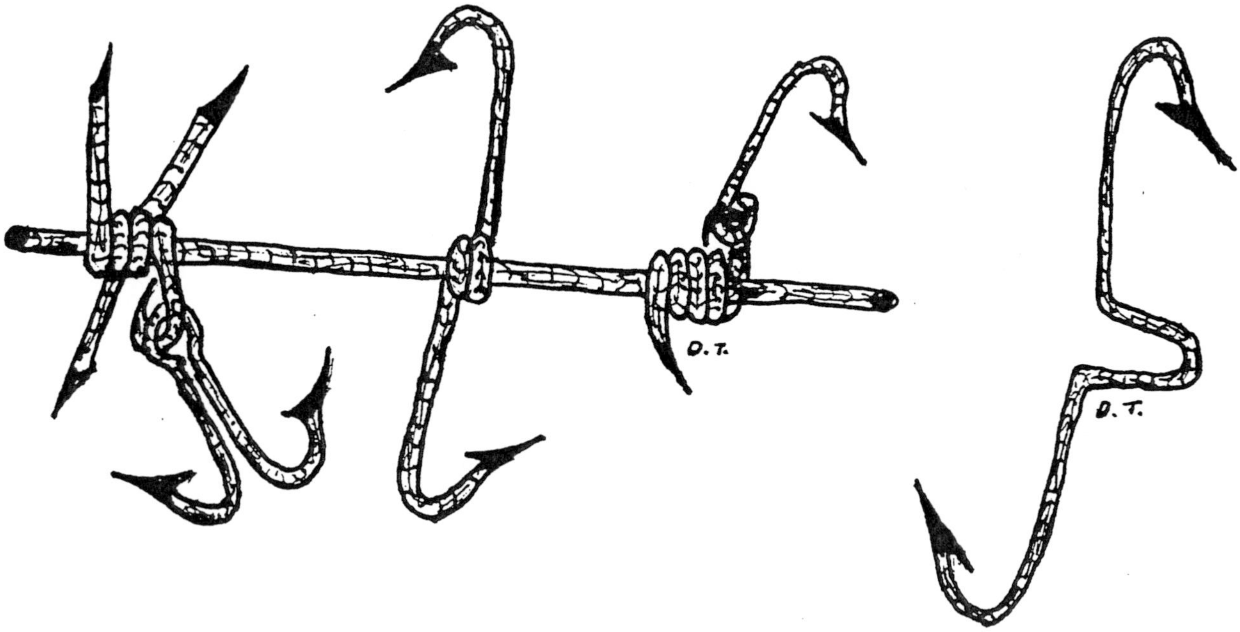


PAT.# 181,608

Patented Aug.29, 1876 by Andrew Upham. Fish hook barb is probably a machine error.

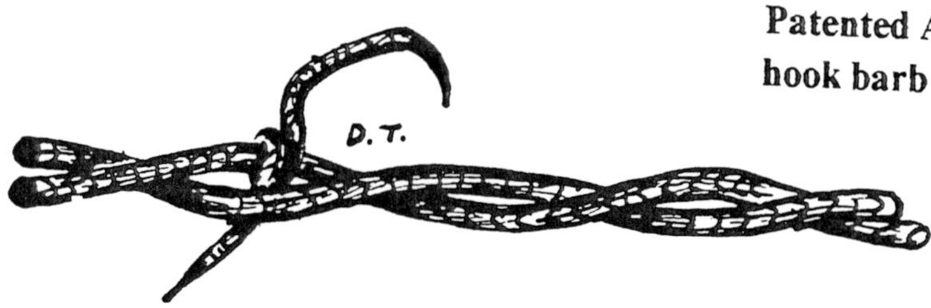


SPIRALING RIBBON WIRE

Patent # and patentee unknown. Made of spring steel wire with fish hook type barbs.

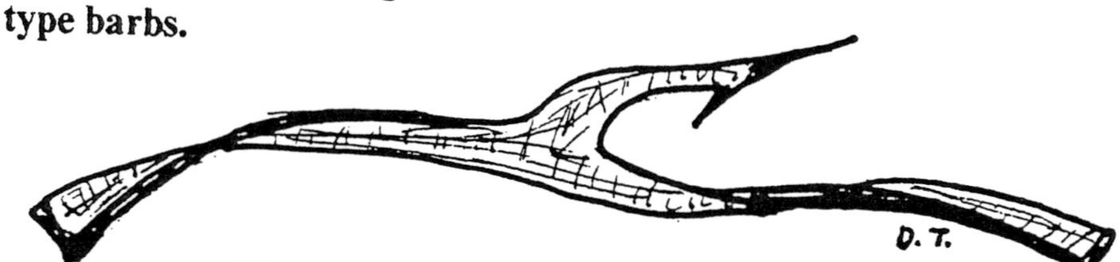

PAT.#2,308,905

Patented Jan.19, 1943 by Harold Abbott. A single strand kinked to hold barbs in place. Barbs may have short loops, long loops, or tight twist design.

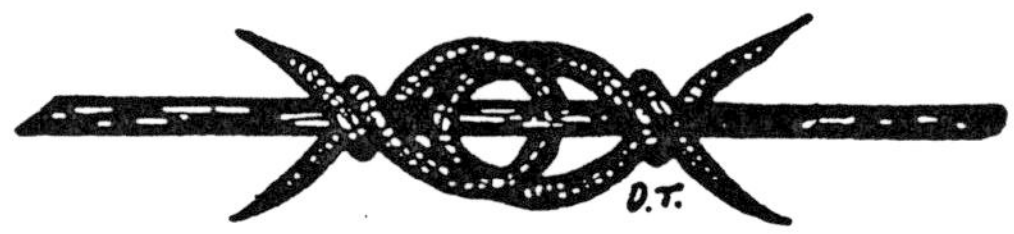

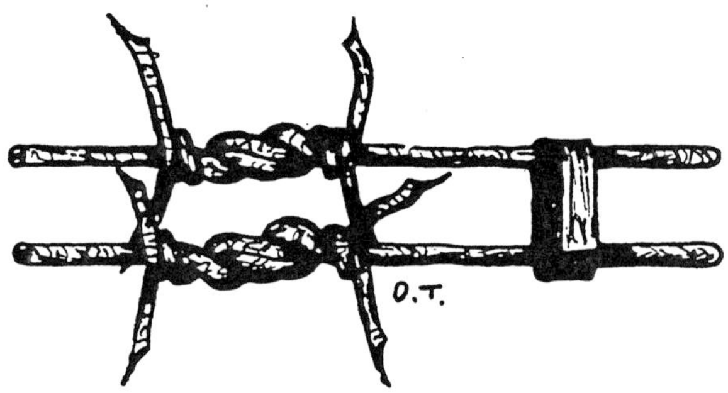


Abbott's Splicing Clip
Used to station wires into a parallel position.

PAT.#2,973,185

Patented Feb.28, 1961 by Rudolf Brixner. An extra large two strand, four point wire used in the Korean conflict.

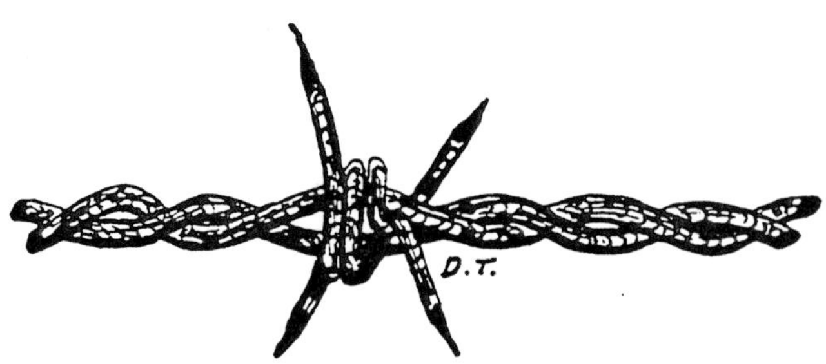

## ITALIAN WARWIRE

**These designs are variations and combinations of gauge sizes and shapes. Names and patents are unknown. Courtesy of an Italian collector.**

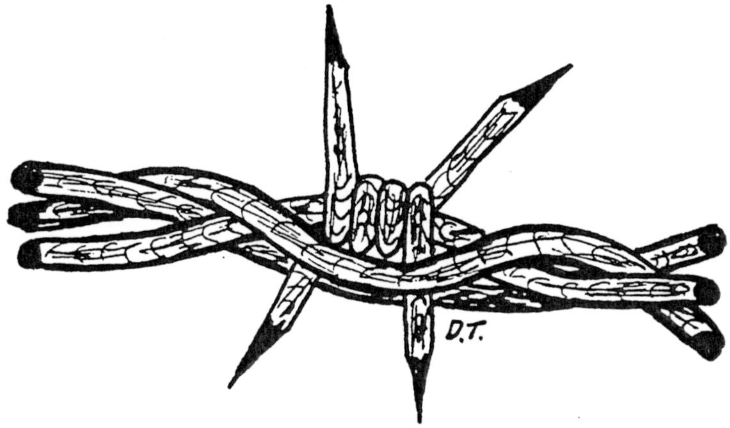


P. AMEDEO-ITALY

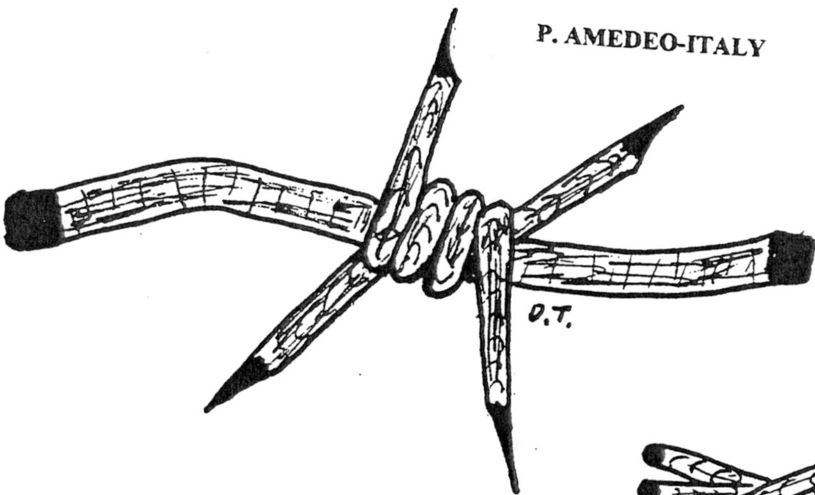


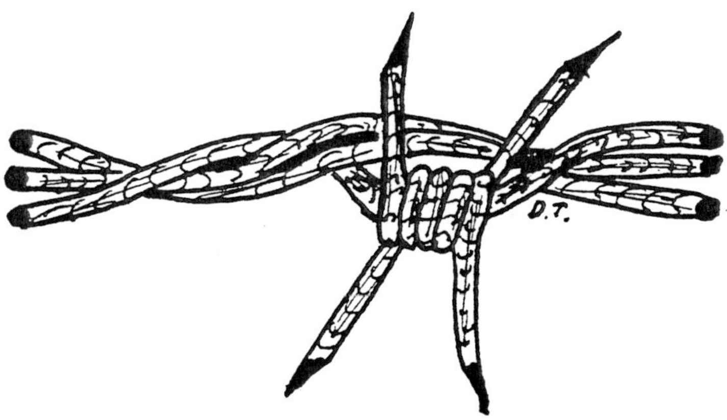


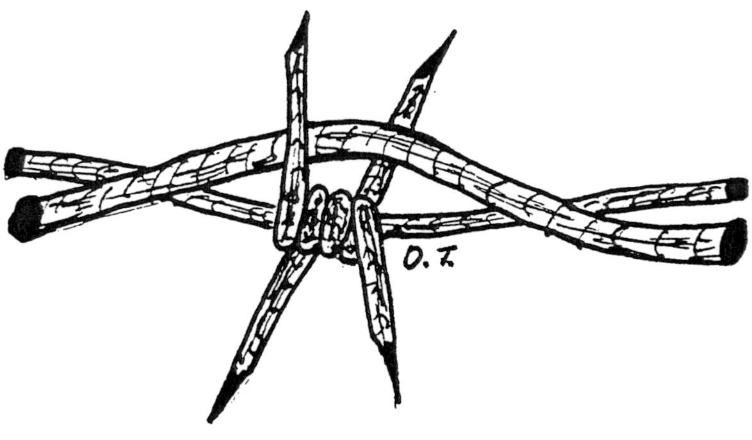

PAT.#6913(reissue-variation)

**Patented Feb. 8, 1876 by Joseph Glidden. A two strand, twisted wire with two point hanging barbs at two and one half inch intervals.**

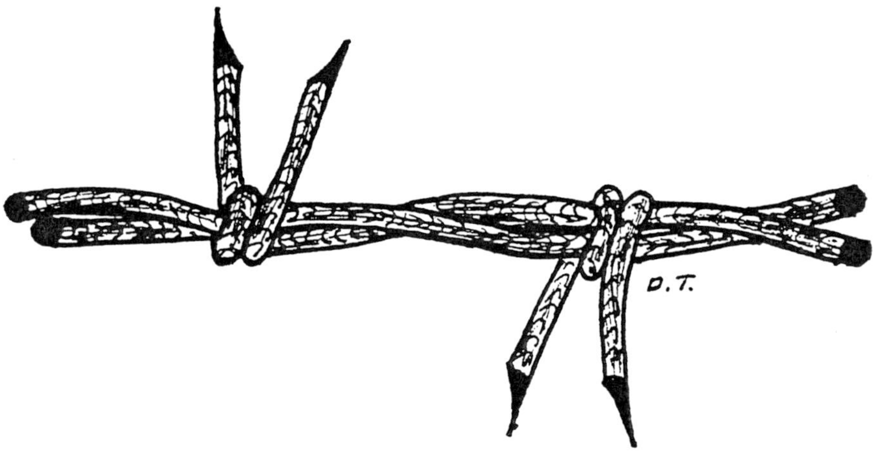


WWII ENGLISH OBSTACLE WIRE

**A single strand, four point barb with extra long barb twist. May be clipped together in both double and triple parallel configurations. Patent # and patentee unknown.**

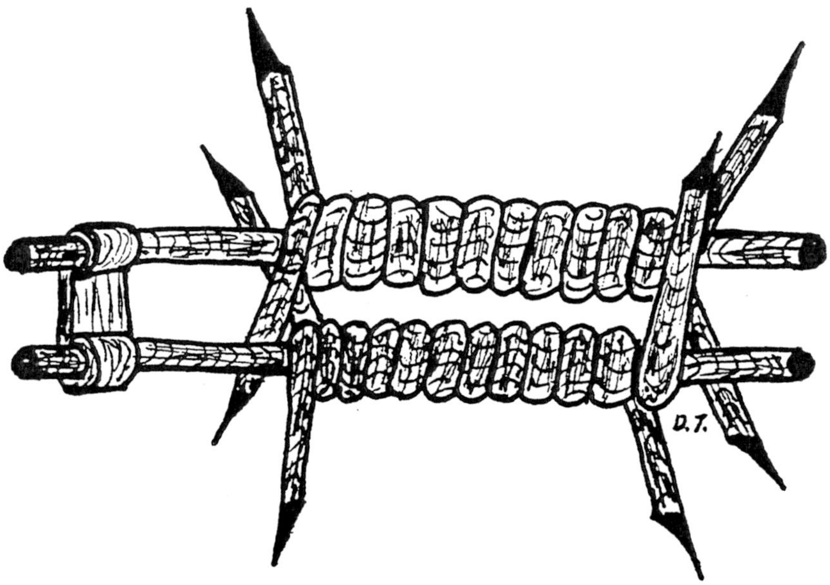

## U.S. EXPERIMENTAL WIRE

Patent # and patentee unknown. This multi-tooth twisted, single strand was probably never produced in significant quantities.

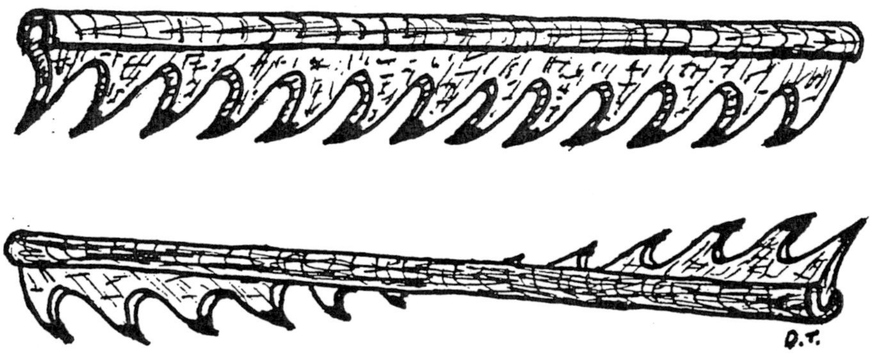


## PAT.#6914(reissue-variation)

Patented Feb.8, 1876 by Joseph Glidden. A single round strand with four point barbs. Spacing of barbs was usually two and one half inches.

## PAT.#6914(reissue-variation)

Patented Feb.8, 1876 by Joseph Glidden. A single, square, twisted strand with four point barbs.

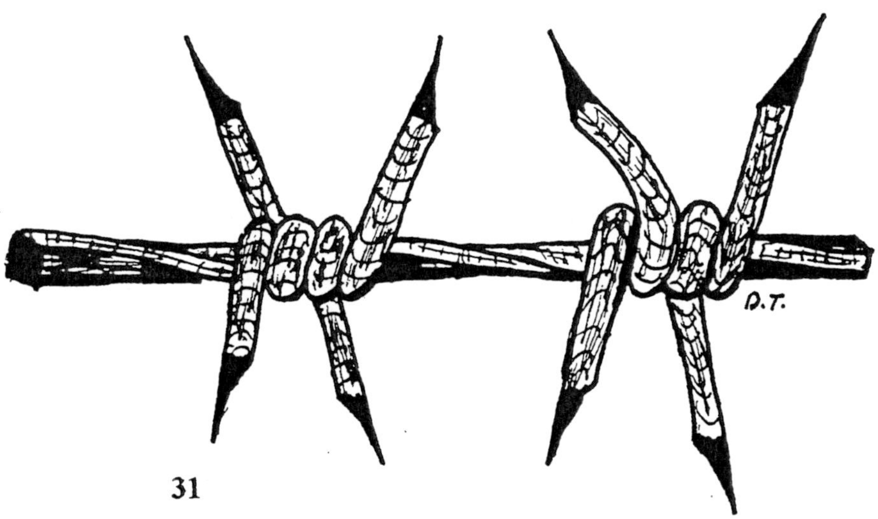

PAT.#3,224,736

**Patented Dec.21, 1965 by Daniel Musgrave. Sheet mesh wire cut into various sizes and shapes to form razor edge obstacle wire. May be flat or twisted.**

PAT.#2,973,185(wire #4)

**Patented Feb. 28, 1961 by Rudolph Brixner. An aluminum version, paired by clip rings, and used temporarily in riot and compound control.**

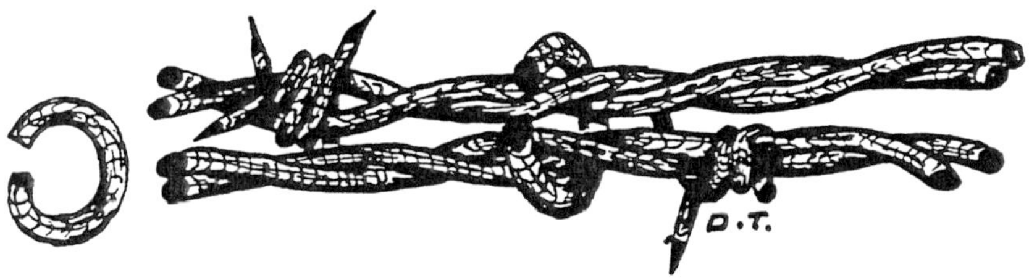


PRE 1900 MILITARY WIRE

**A single strand, one point barb, patent # and patentee unknown.**

PAT.#3,455,539
**Patented July 15, 1969 by J.G. Loofbourrow. A flat ribbon wire with vicious fish hook barbs.**

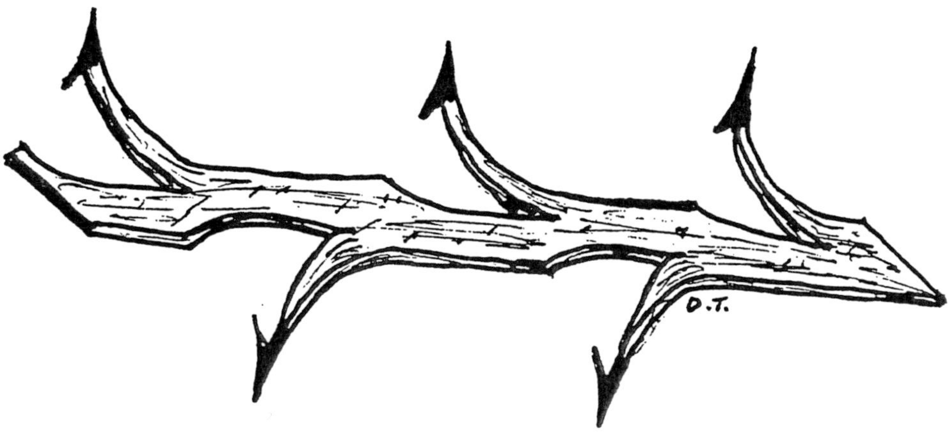


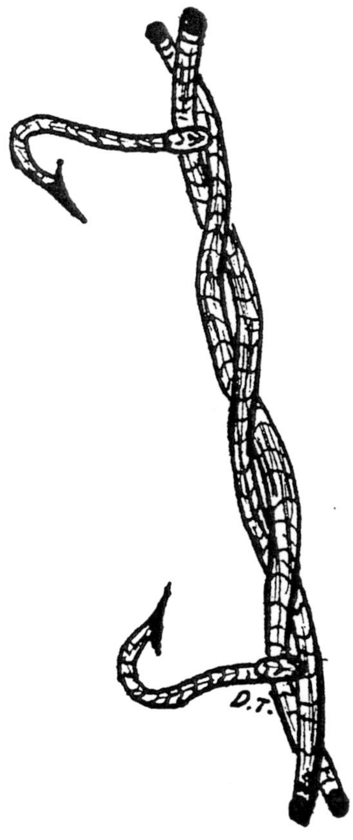


PAT.#1,142,126
**Patented June 8, 1915 by Thomas Sturtevant. Various designs of other fish hook type wire.**

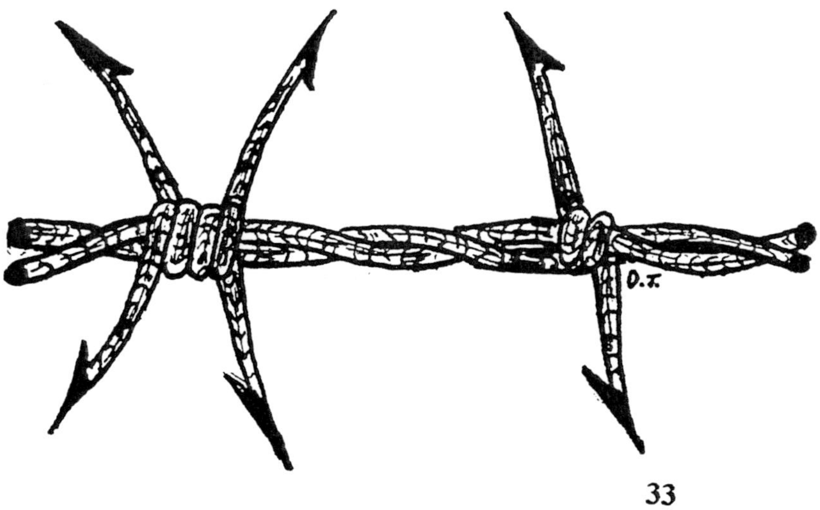

PAT.#4,328,955

Patented May 11, 1982 by Albert Hermans. A vicious barbed ribbon wire used by the military and used as security wire by the private sector.

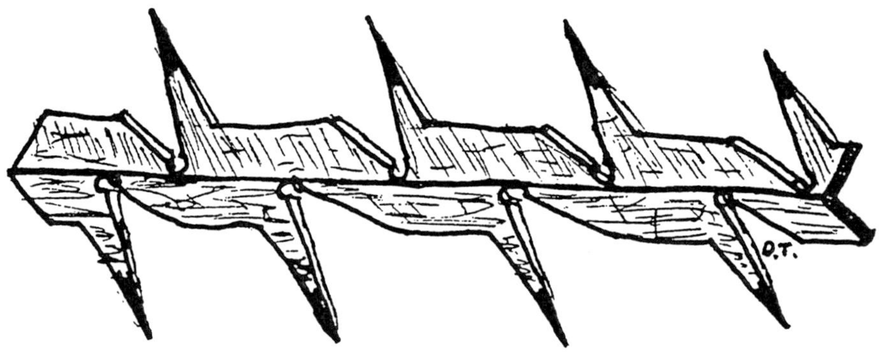


PAT.#3,615,273

Patented Oct.26, 1971 by Jacques See. A flat tape cut in a fashion to extend barbs when unrolled.

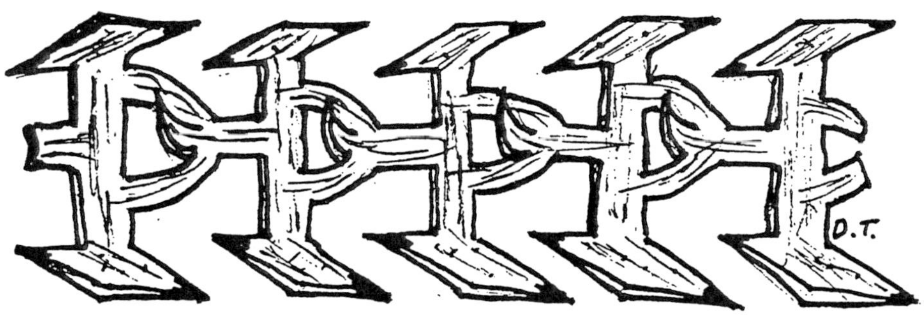


PAT.#2,915,104

Patented Dec. 1, 1959 by George Rasmussen. A part of a machine patent making ribbon wire.

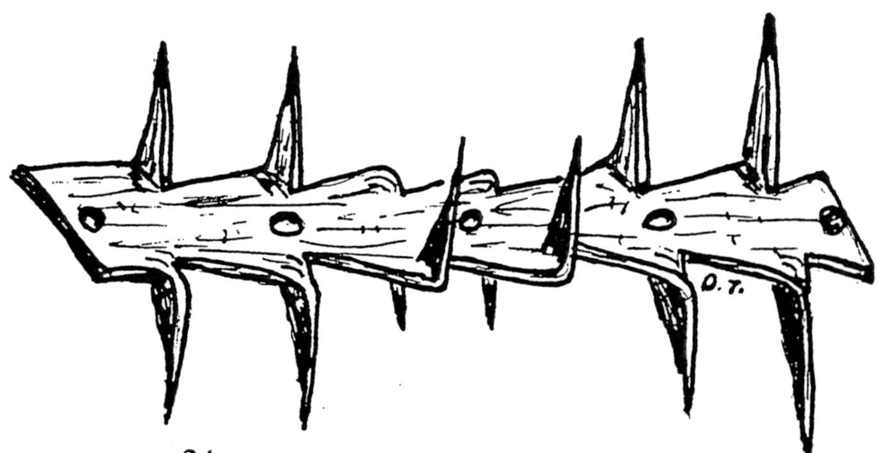

**PAT.#2,908,132**

**Patented Oct. 13, 1959 by Ernst Klemm. A flat, sharp razor wire made in Germany.**

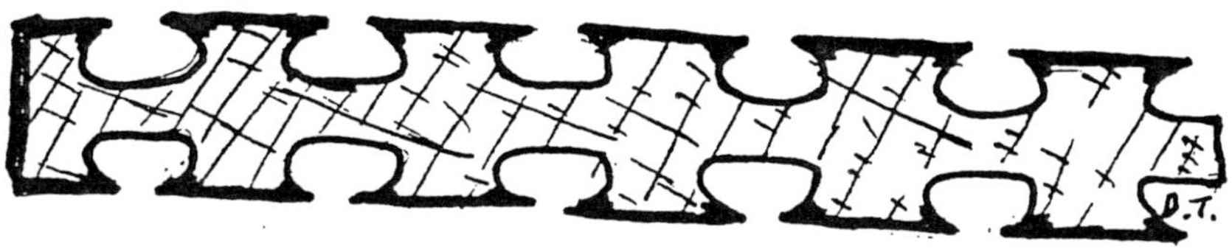


**PAT.#2,908,484**

**Patented Oct.13, 1959 by Siegfried Uhl. A tough razor wire with a center bead and offset opposed barbs.**

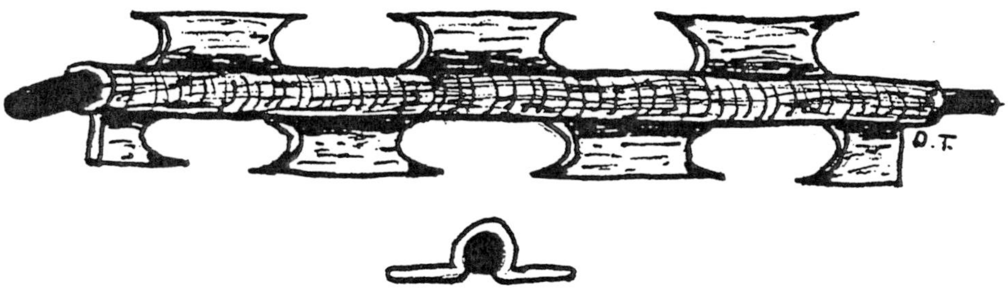


**PAT.#3,941,164**

**Patented Mar. 2, 1976 by D. Musgrave. A flat or twisted steel tape with razor type barbs.**

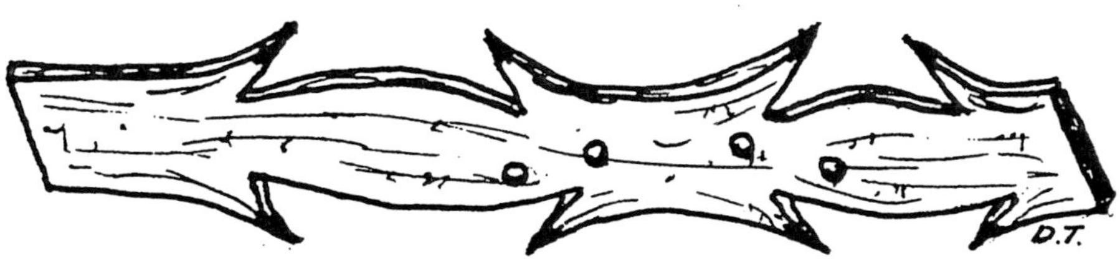

**PAT.#4,028,925(machine)**

Patent # and patentee unknown. Origin shown under the above machine patent. Used in Vietnam and made in two widths. No center bead, holes made for tying .

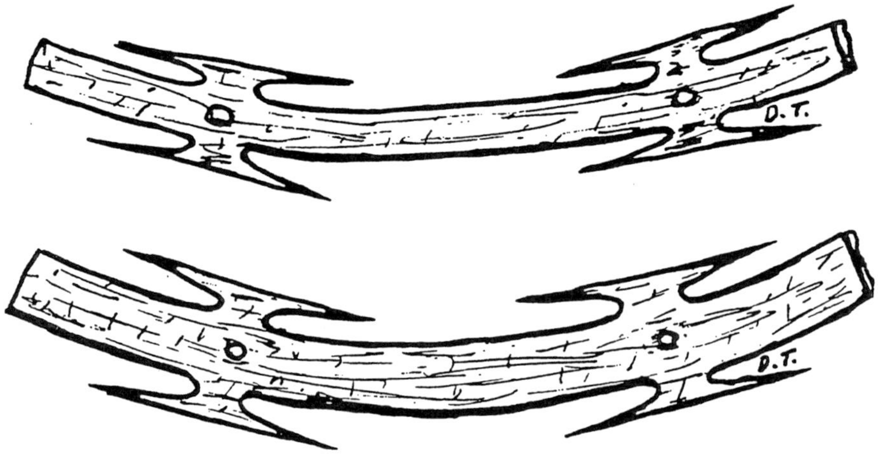


**PAT.#3,463,455**

Patented Aug.26, 1969 by Paul Meckel. A vicious coiled ribbon with double razor points.

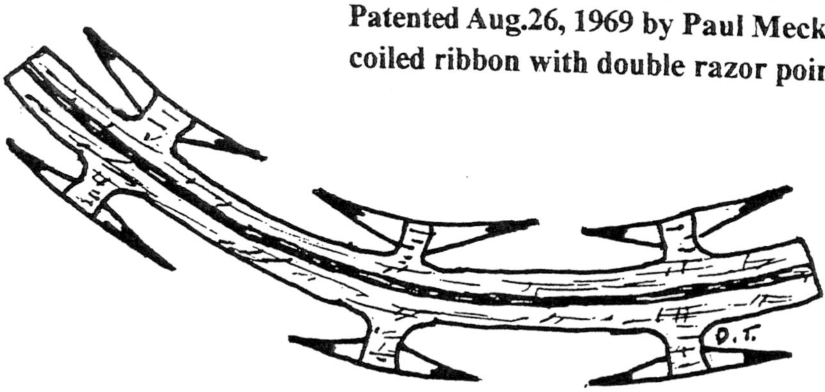


**PAT.#4,028,925**

Patented June 14, 1977 by Michael Mainiero. A razor wire pre-bent into coils. Shown in a machine patent drawing.

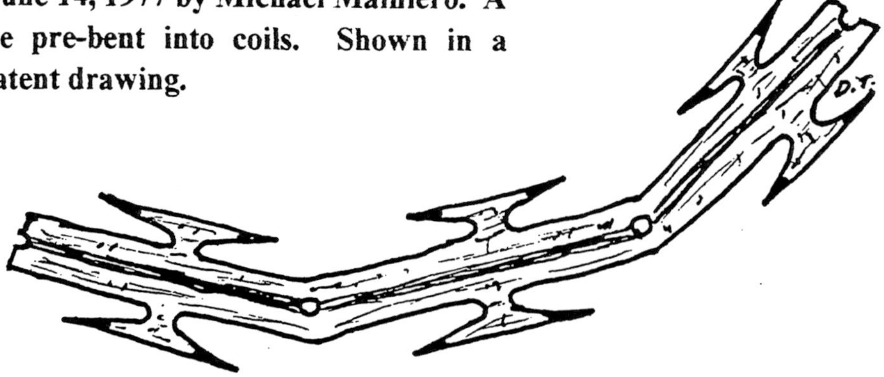

PAT.#3,990,485

Patented Nov. 9, 1976 by Siegfried Uhl. A razor wire with a beaded center allowing from two to four lines of barbs to extend outwards.

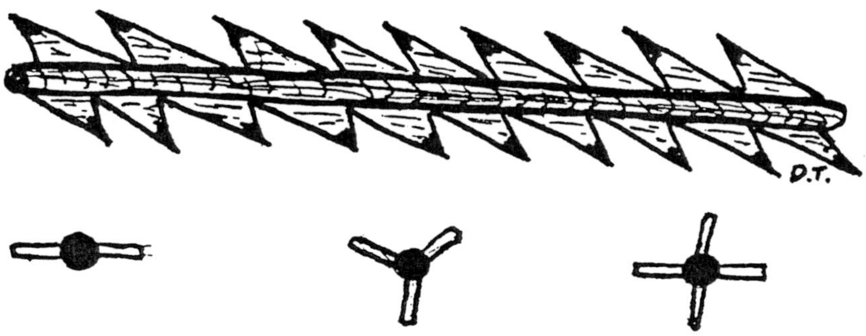


PAT.#2,392,223

Patented Jan.1, 1946 by Alan Brickman. Shown is a single strand of a barbed wire fabric or mesh that can be cut or formed for various uses.

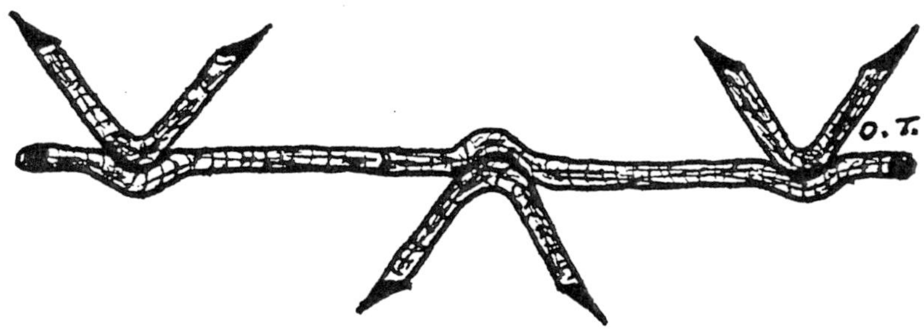


PAT.#3,010,701

Patented Nov. 28, 1961 by Ernst Klemm. A spring steel razor wire, pre-tensioned to twist after being unrolled for use. Made in Germany.

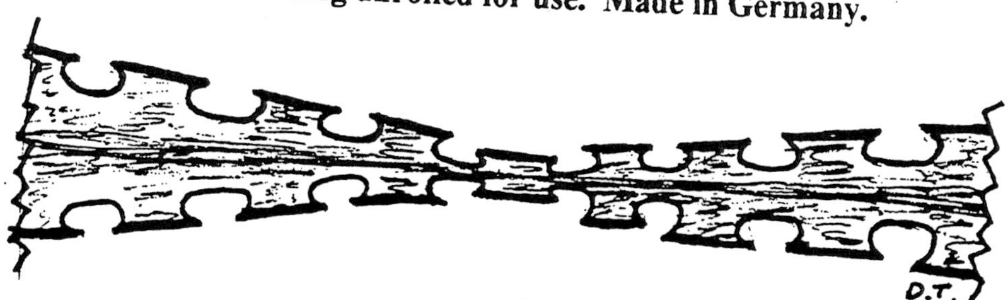

**FRENCH MILITARY WIRE**

**Two strand, four razor point, patent # and patentee unknown.**

**PAT.#3,763,529**

**Patented Oct. 9, 1973 by Daniel Musgrave. A flat or twisted vicious razor wire with a beaded center. Almost impossible to cut through.**

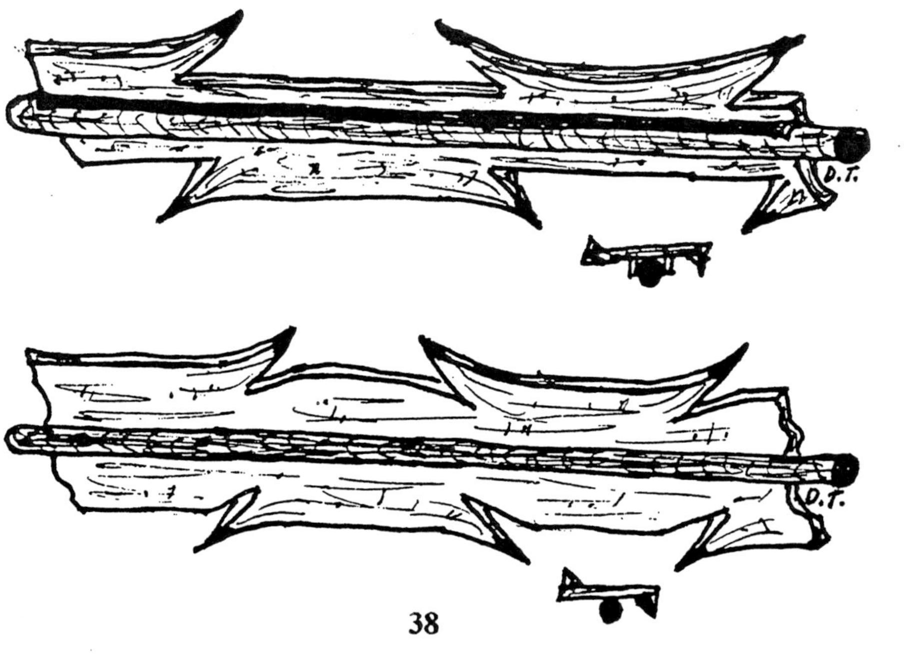

**PAT.#252,746(variation)**
**Patented Jan.24, 1882 by Dodge & Washburn. A two strand, four point wire with extra long barbs.**

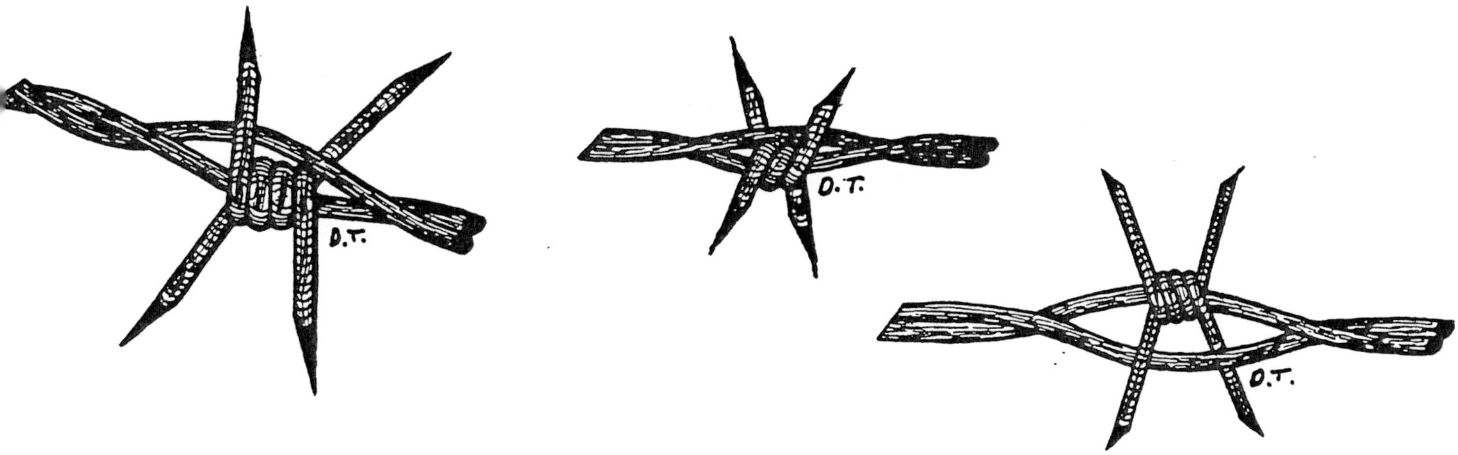


**PAT.#4,031,284**
**Patented June 21, 1977 by Glen Ingraham. An anti-intrusion chain link fence with barbs at the top, acting as a ground for an electrically charged hot wire at the top.**

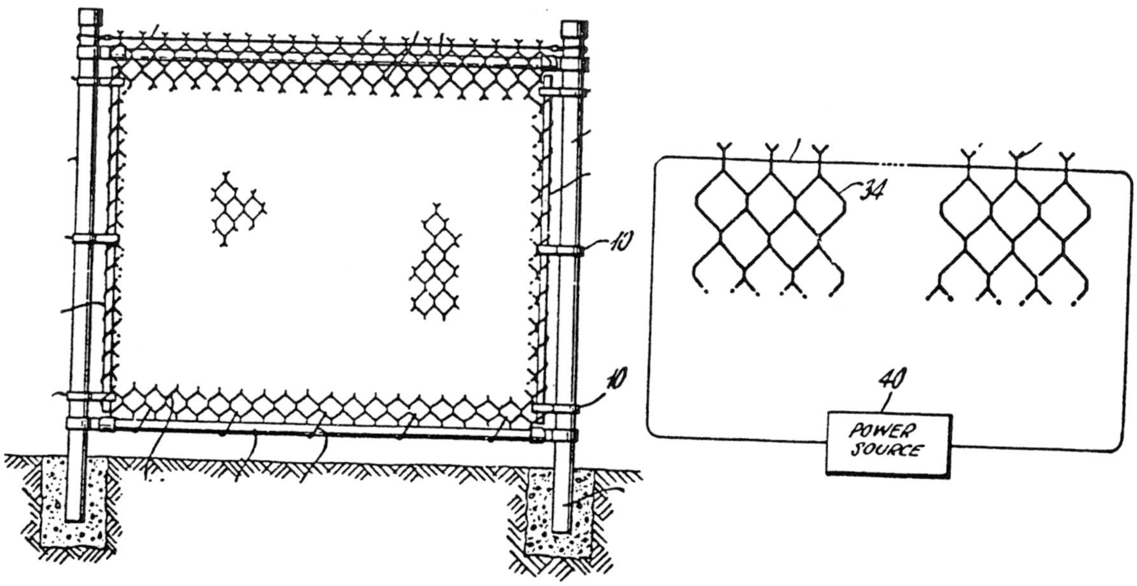

## SUGGESTED FOR FURTHER STUDY

Barriers by Campbell & Allison

Barbs, Prongs, Points, Prickers, & Stickers by Robert Clifton

Bobbed Wire Bibles- Vol. I -IX by Jack Glover

Early U.S. Barbed Wire Patents by Jesse James

Old West Barbed Wire & Patent Dates by Jay Jenkins

History Of The Manufacture Of Barbed Wire by Boone McClure

The Barbed Wire Collector Magazine by DRM Publishing Co.

# SECTION II.

## WARWIRE TOOLS AND DEVICES

**Patented devices invented to combat the entanglement wire barriers used in warfare.**

# INTRODUCTION TO SECTION II.

As use of entanglement wire grew, so did the development of special tools with which to work the wire. The troops at Vicksburg broke the telegraph wire by hand and entwined it among existing tree stumps and posts. No tools were needed. Later, the French used smooth wire and wooden posts which required few tools.

When barbed wire came onto the scene, wire cutters and leather gloves became a necessity. No records of military issue tools for working wire have surfaced to date, so the regular stockmen type tools were probably used.

Iron posts began to appear in the early 1990's, as stockmen over the world built barbed wire fences. Photographs and paintings of both the Boer War and the Russo-Japanese War show a variety of wooden and iron posts used in entanglements. The Spanish American War photographs show only wooden posts and conventional fencing used.

One post invention, a heavy iron rod with a corkscrew bottom and preformed eyelets above, proved popular with the military. No special tools were needed to twist the post into the soil and attach wire to the eyelets. It could also be recovered easily for reuse if needed.

Basically, military tools used for building entanglements was an exercise in adaptation of stockmen tools. This was not the case with tools needed by the aggressor. Barbed wire entanglements quickly became the most formidable obstacle encountered by battlefield troops.

To meet this challenge, military engineers and inventors from the world over, presented a multitude of ideas and patents to combat barbed wire barriers. Some of these ideas are presented in this section.

There were two fields of thought. Either equip the individual soldier so he has the capability to cut his way through the wire, or develop special equipment capable of going through the wire in a way that troops could follow.

Since progress by individuals was slow and dangerous, efforts turned to building equipment like the steel tracked tank, specially designed explosives to penetrate the entanglements, and bombardment tactics whereby the defensive positions were rendered useless. The tools of war changed to meet the needs.

## MODERN BAYONET/WIRE CUTTERS

RUSSIAN—1891-1930 REGULAR SLICK MODEL

J. JANZEN

SAME AS ABOVE EXCEPT WITH "WIRE BREAKER" INSTALLED

UNITED STATES—M9 1989–This bayonet/wire cutter was developed by the military when it was discovered U.S. soldiers were carrying their own survival knives to the field. This model replaces model carried for 25 years, and weighs 1.8 lbs. Knife used in combination with the scabbard for wire cutting.

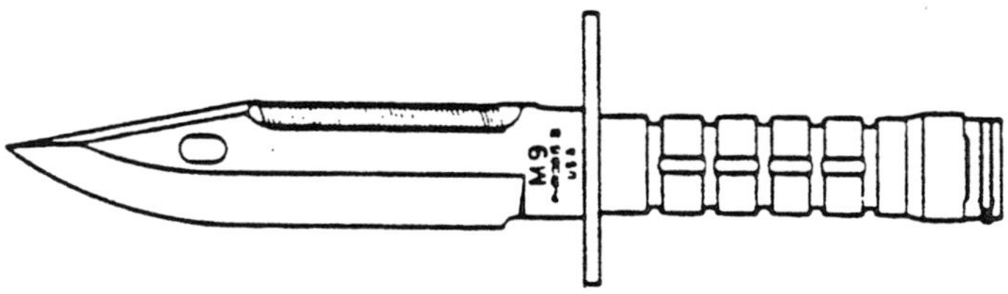

J. JANZEN

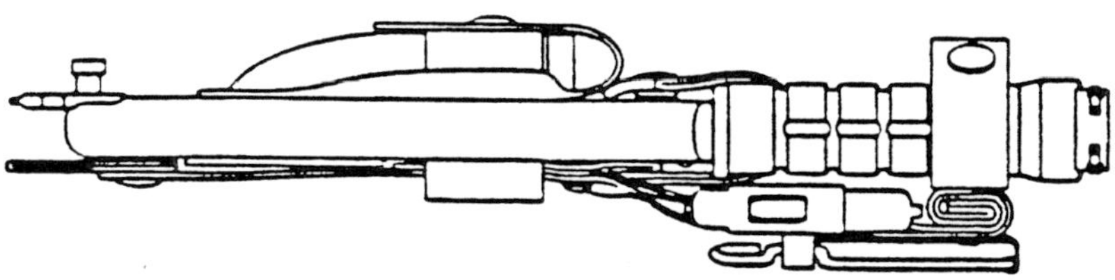

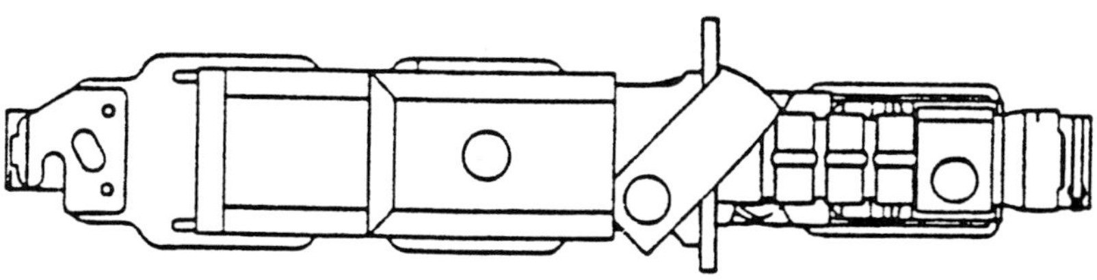

PAT.#1,253,627
Patented Jan.2, 1918 by W.T. O'neill
A combination bayonet/wire cutter designed to sever wire by thrusting the rifle forward.

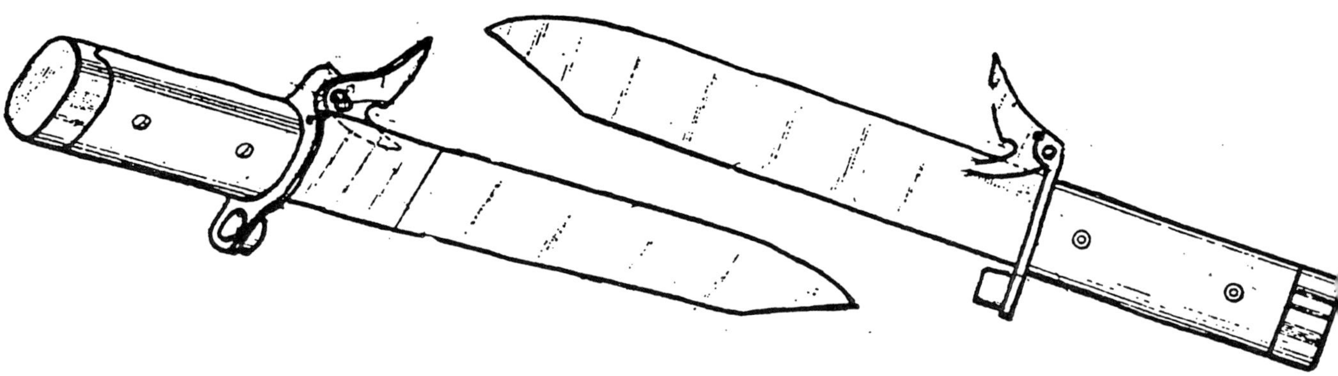

PAT.#1,274,019
Patented July 30, 1918 by W. Dziezyc
A combination bayonet/wire cutter designed to be operated by pull rope allowing operator to remain in a safe position under fire.

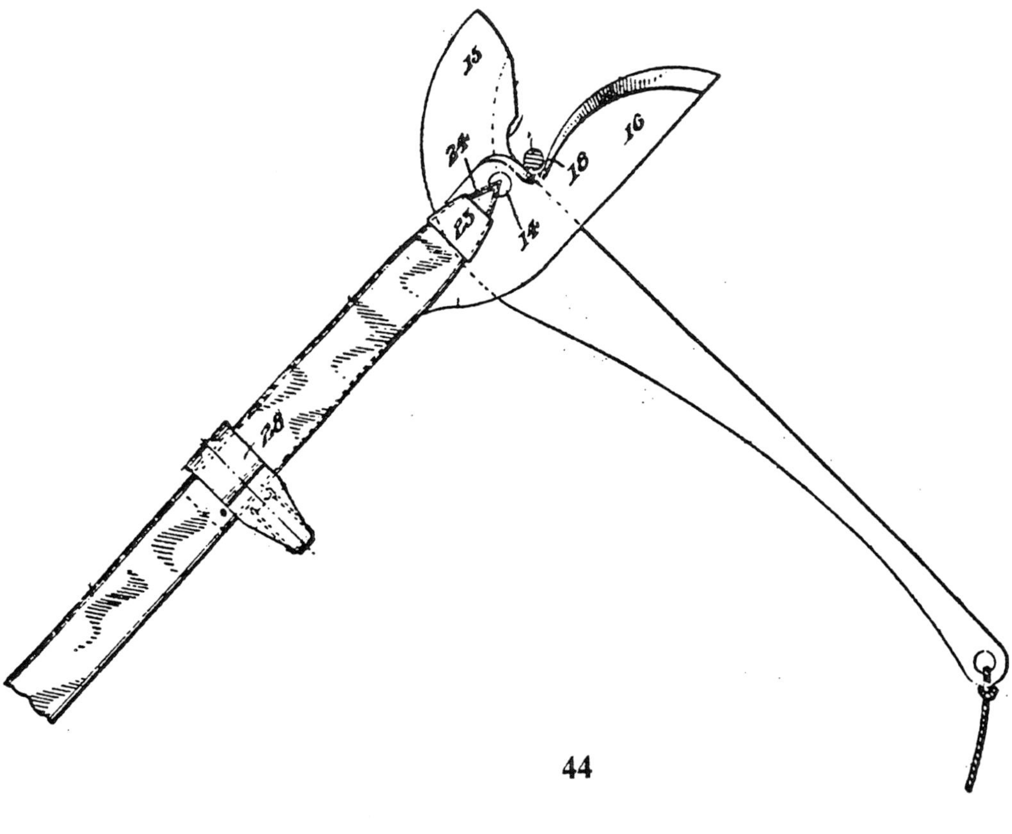

## MODERN BAYONET/WIRE CUTTER COMBINATIONS
## CIRCA 1940-1990

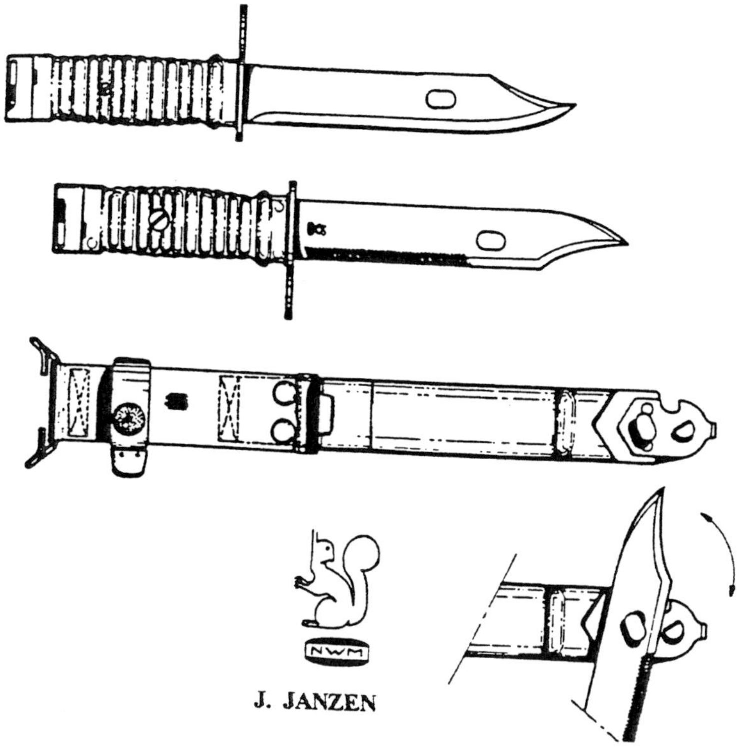


J. JANZEN

NEDERLANDS—STONER BAYONET/WIRE CUTTER
used in combination with the scabbard.

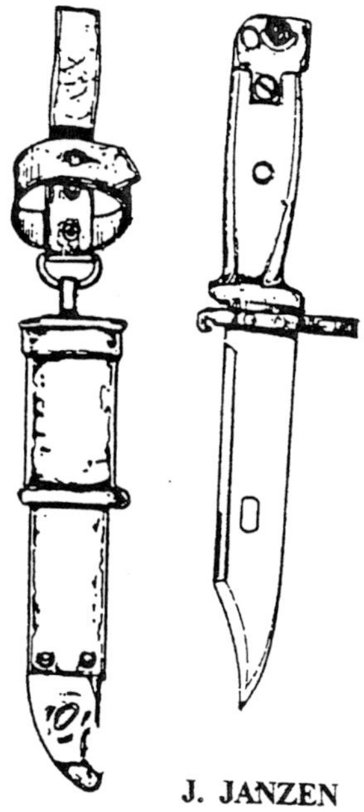

J. JANZEN

POLISH—MODEL 1978
BAYONET/WIRE CUTTER
used in combination with
the scabbard.

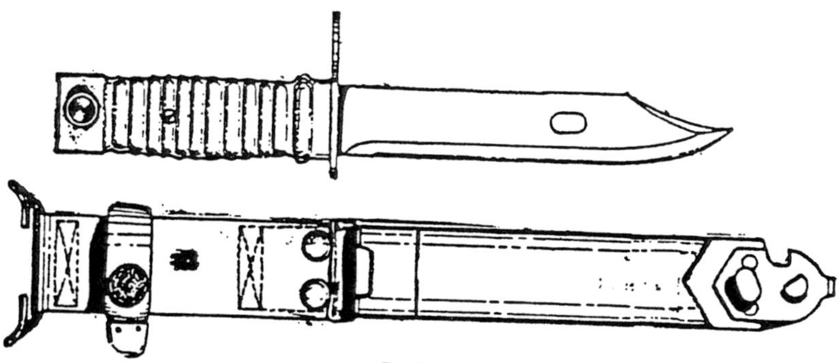

J. JANZEN

SWISS—E.W.–1970 BAYONET/WIRE CUTTER
used in combination with scabbard.

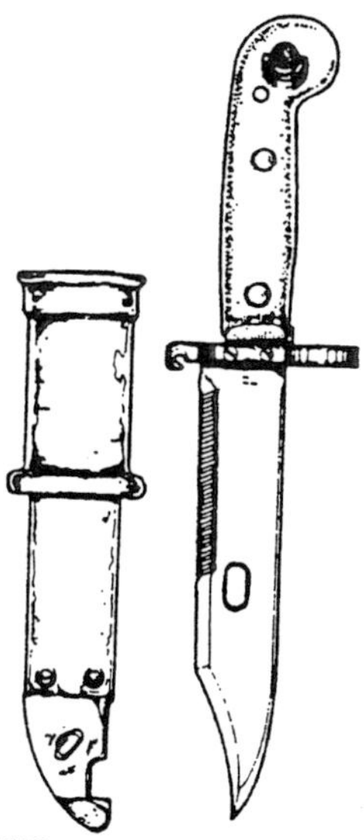

J. JANZEN

RUMANIAN—AKM BAYONET/WIRE CUTTER
used in combination with scabbard.

## MODERN BAYONET/WIRE CUTTERS

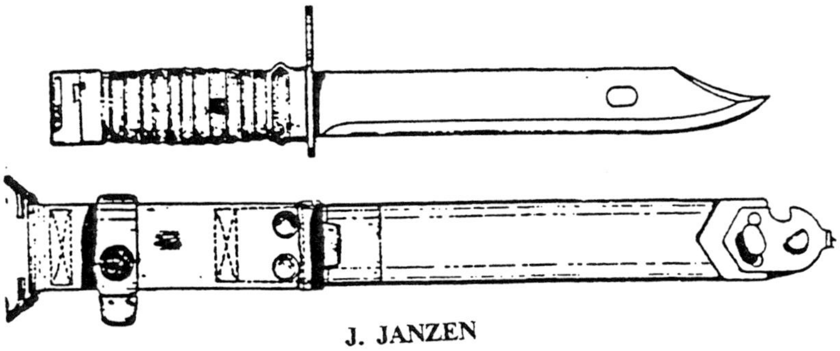

J. JANZEN

**DUTCH—KCB-77LM1-1977 BAYONET/WIRE CUTTER**
**used in combination with scabbard.**

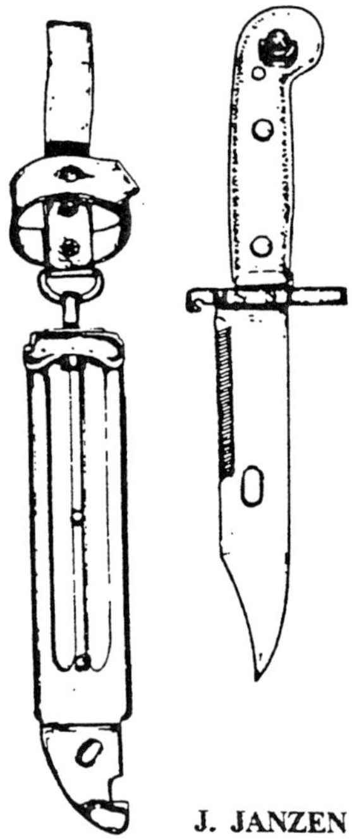

J. JANZEN

**EAST GERMAN—AKM BAYONET/WIRE CUTTER**
**used in combination with scabbard.**

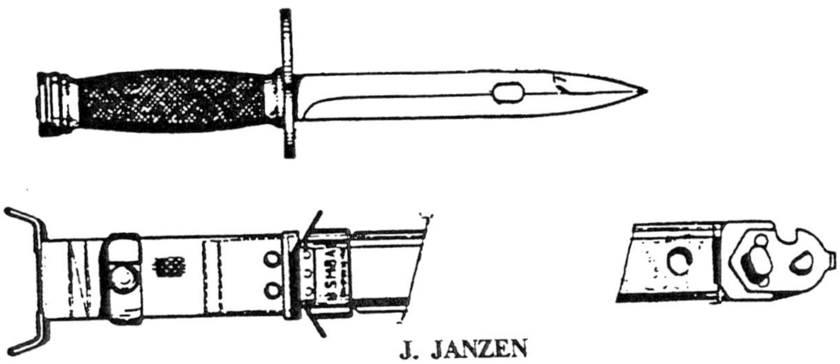

J. JANZEN

**UNITED STATES—M7 BAYONET/WIRE CUTTER**
**Made in Germany-used in combination with scabbard.**

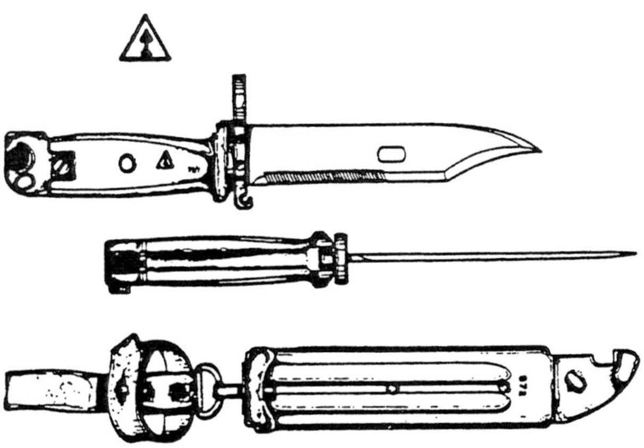

J. JANZEN

**RUSSIAN—AKM 1972 BAYONET/WIRE CUTTER**
**Fore runner of modern day models**
**used in combination with scabbard.**

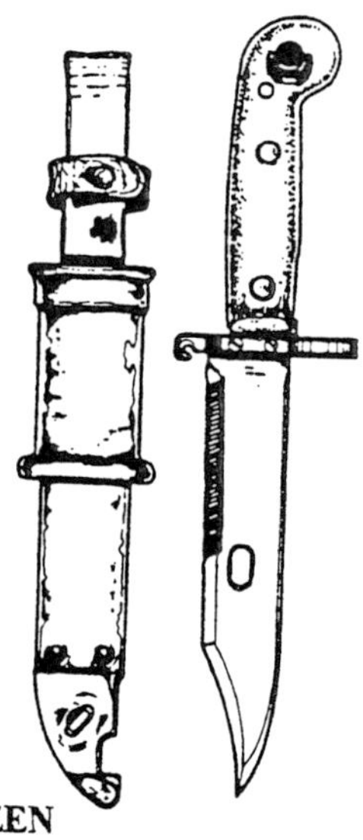

J. JANZEN

**RUSSIAN—SVD 1968 INSULATED FOR CUTTING ELECTRICAL WIRES. Used in combination with insulated scabbard.**

**PAT.#1,260,723**
**Patented Mar. 26, 1918 by J. Skwarek**
**A combination bayonet/wire cutter/scissor device to be used in cutting barriers in warfare.**

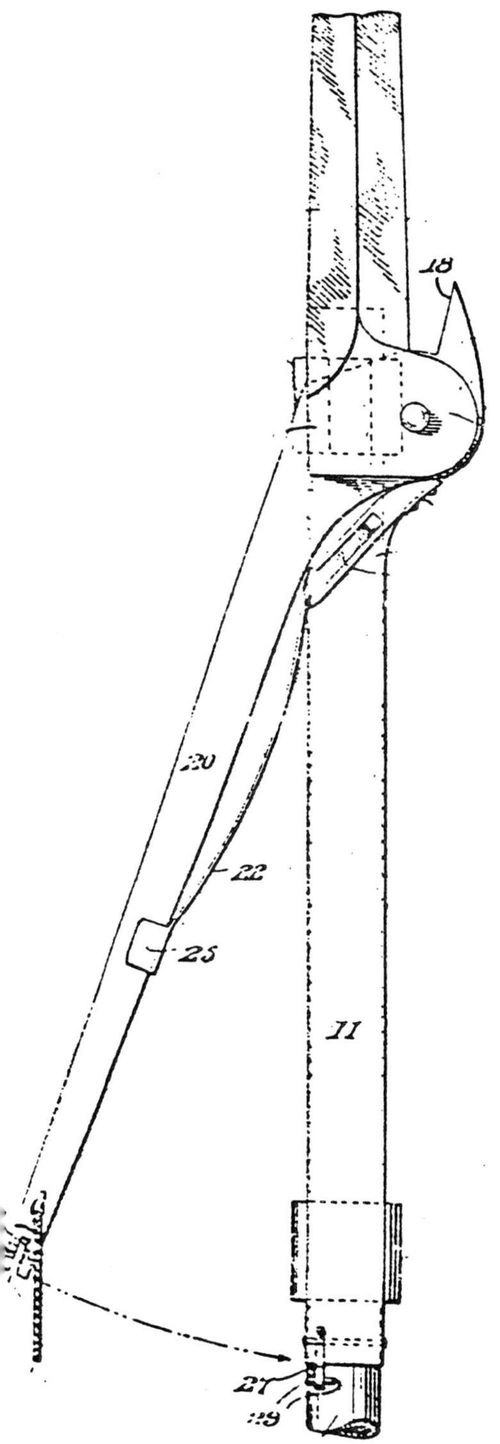


**PAT.#822,441**
**Patented June 5, 1906 by T.S. Forbes**
**A combination bayonet/wire cutter/pliers designed for both cutting and gripping. Thereby avoiding the necessity of providing additional equipment.**

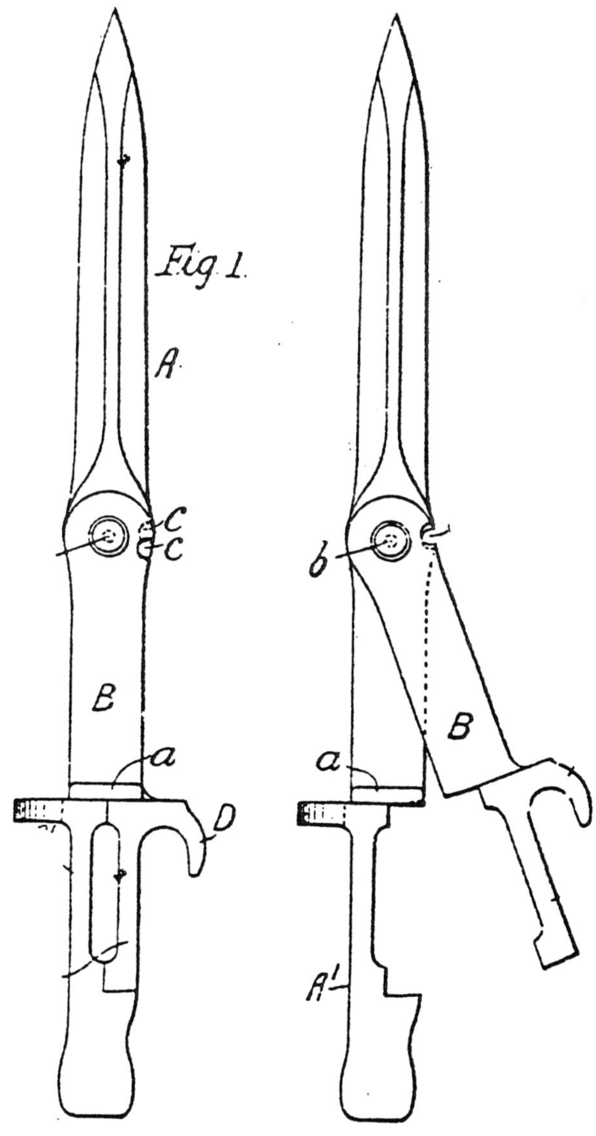

**PAT.#1,694,753**
**Patented Dec. 11, 1928 by J. Peppin**
**A portable barrier easily moved with troops.**

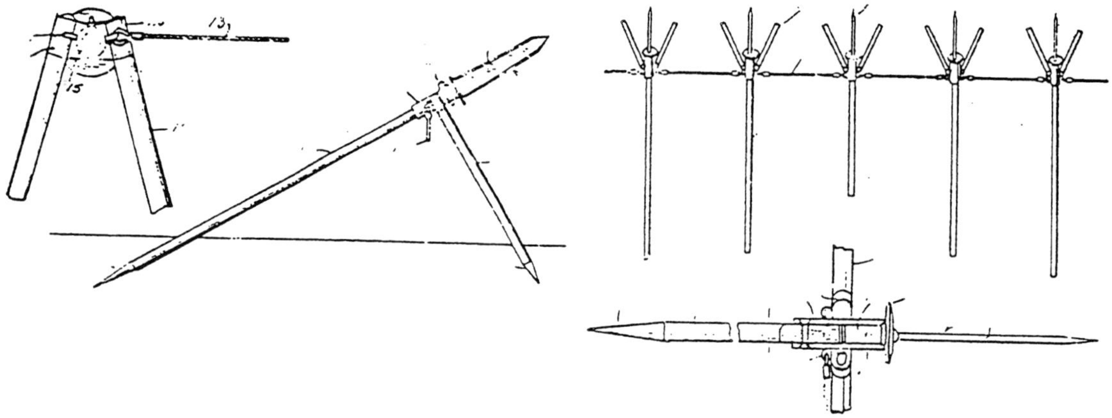

**PAT.#338,917**
**Patented Mar. 30, 1886 by J. Bloomer**
**An attachment to existing fences and walls thus preventing instrusion .**

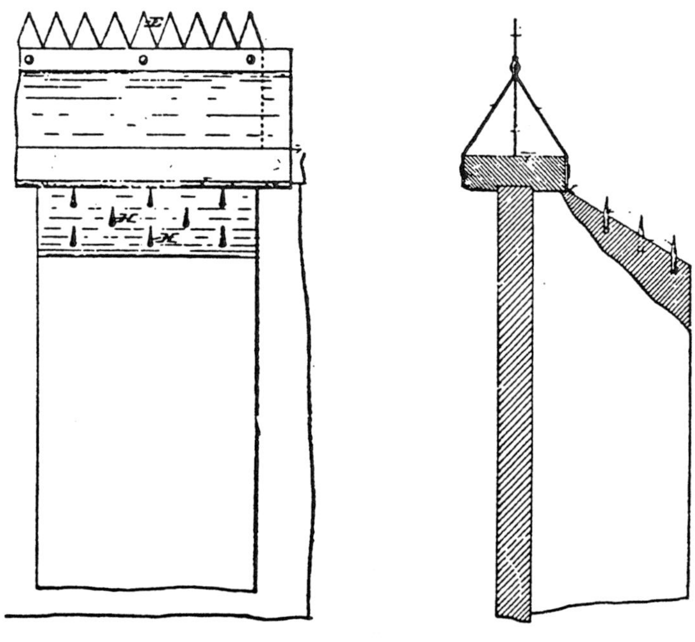

**PAT.#325,749**
**Patented July 10, 1906 by G.H. McMaster**
**A combination wire cutter/entrenchment tool designed to be carried inside the stock of a rifle.**

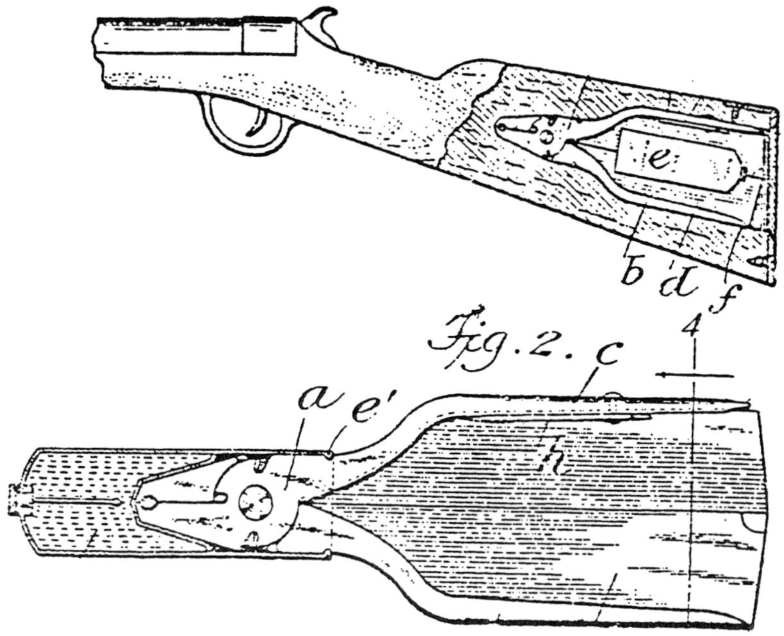


**PAT.#1,290,674**
**Patented Jan.7, 1919 by B. Sorocinsky**
**A lightweight wire cutter attachment that can be slipped over the blade of a knife or spear, and is actuated by thrust action.**

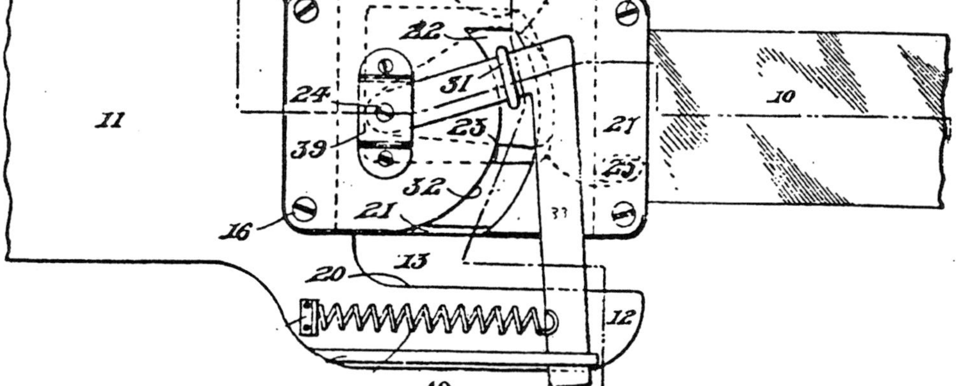

PAT.#883,494
Patented Mar. 31, 1908 by K. Shiramine
A portable alternative to wire barriers, similiar to the caltrop in a larger size.

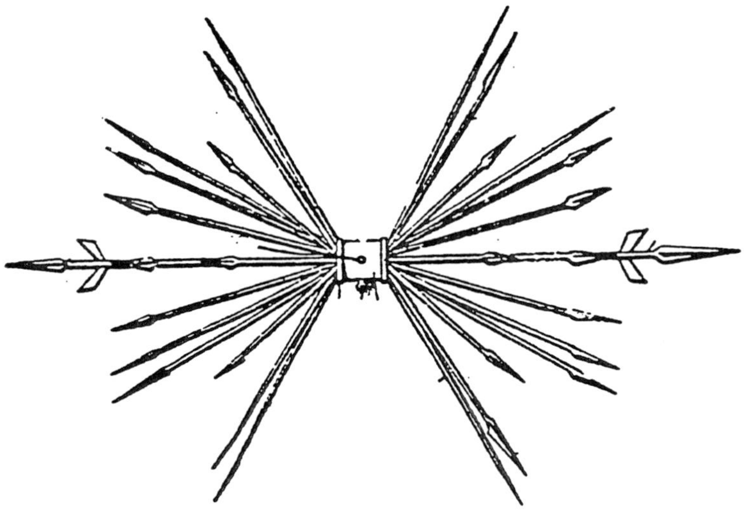

PAT.#1,290,581
Patented Jan. 7, 1919 by P. Kozbial
An attachment for rifle or pole with which to cut entanglement wire from a safe position.

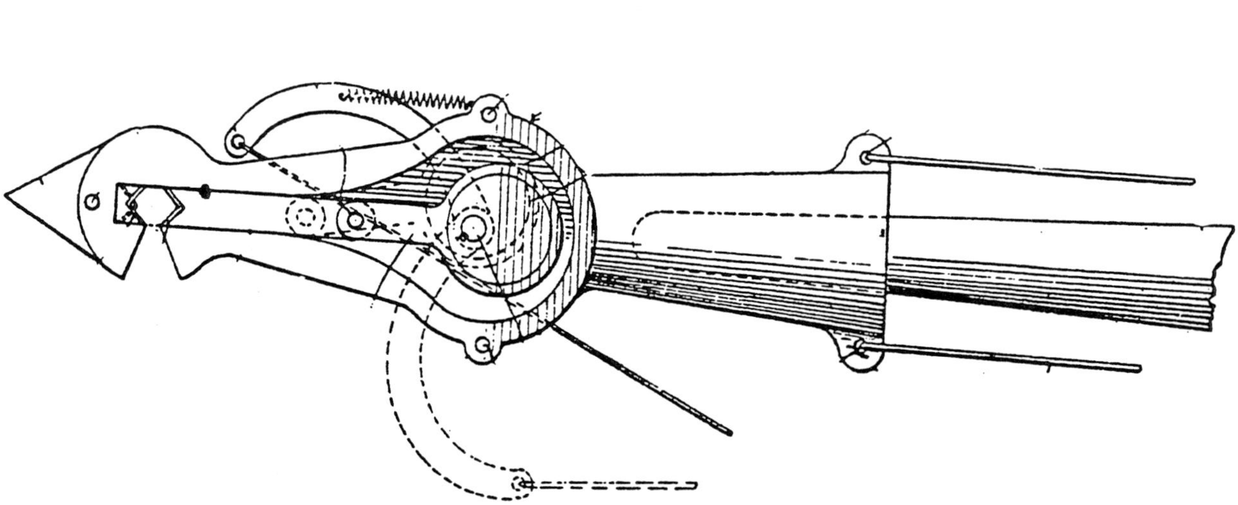

**PAT.#1,269,790**
**Patented June 18, 1918 by J. Danhoffer**
**A combination bayonet/wire cutter designed to allow both rifle mounted and hand use.**

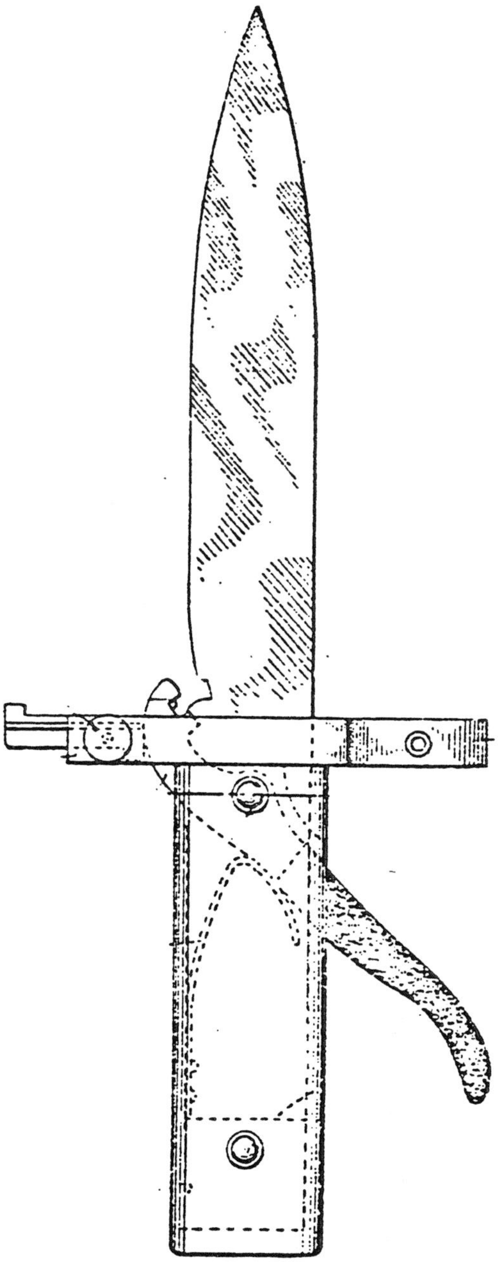

**PAT.#1,269,126**
**Patented June 11, 1918 by M. Stroke**
**A combination bayonet/wire cutter with ease of use yet not hindering operation of rifle and bayonet.**

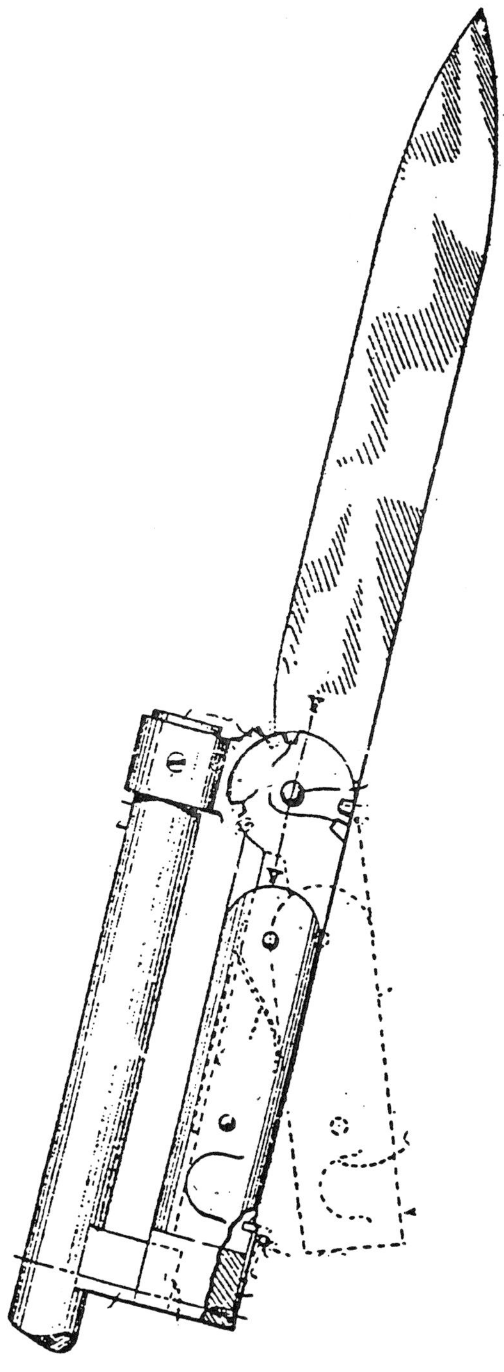

**PAT.#1,268,172**

**Patented June 4, 1918 by J. Stanislawski**

**A combination bayonet/wire cutter designed to be inconspicuous yet ready for instant use.**

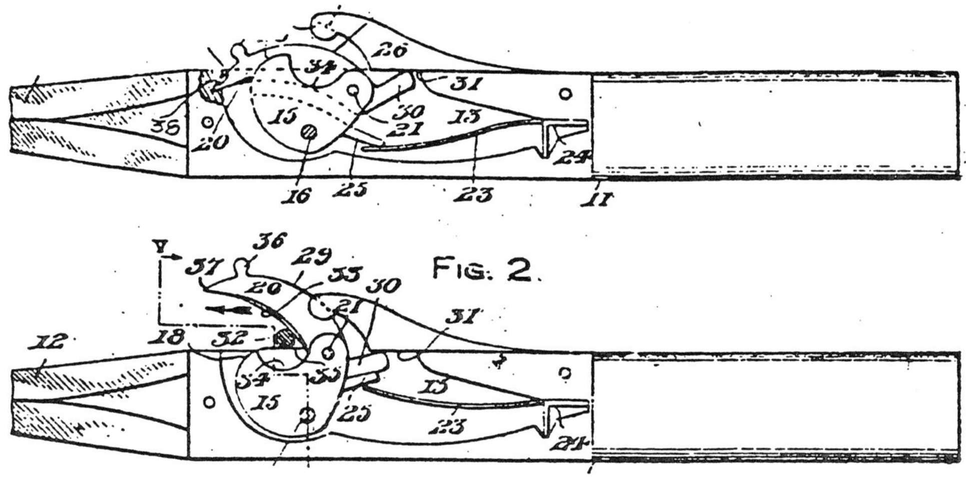


**PAT.#1,266,640**

**Patented May 21, 1918 by C. Szafranski**

**A combination bayonet/wire cutter/shear built for great strength and cutting ability.**

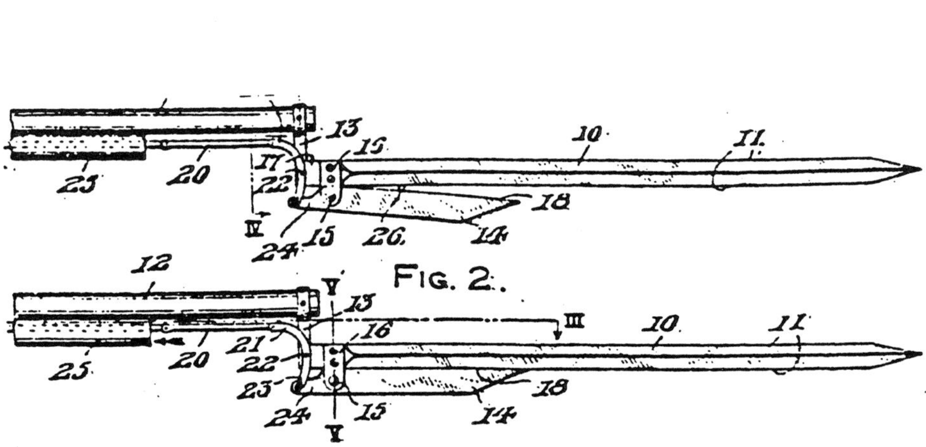

PAT.#1,282,903

Patented Oct. 29, 1918 by P. Maziarz

A combination bayonet/wire cutter offering compact storage, being instantly available, and adding little weight to the weapon.

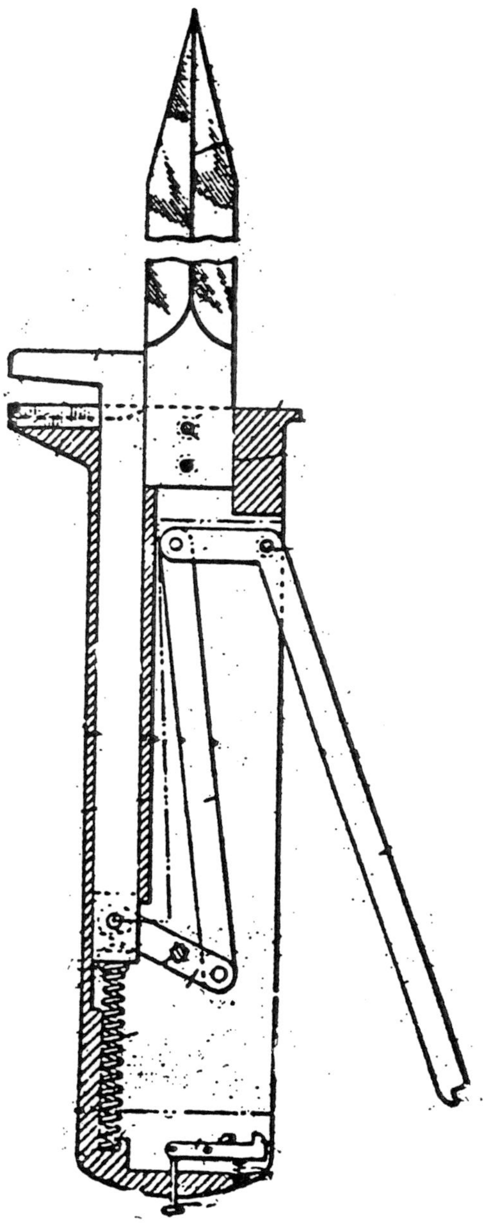

PAT.#1,277,521

Patented Sept. 3, 1918 by J. Worek

A combination bayonet/wire cutter featuring a wire hook that when pushed against the wire, it activates a cutter severing the subject wire.

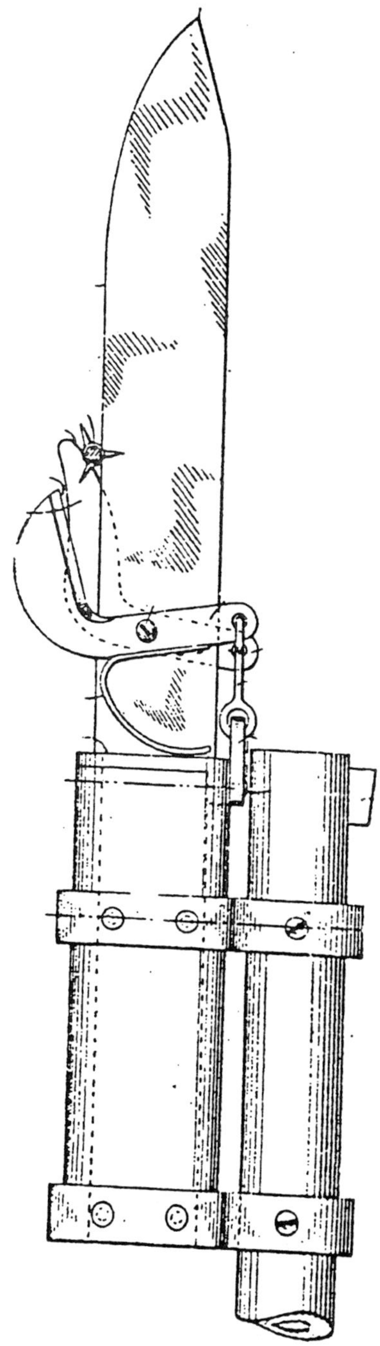

PAT.#1,274,194
Patented July 30, 1918 by J. Pirog
A combination bayonet/wire cutter/shear designed to provide heavy shearing ability yet folds out of the way for bayonet use.

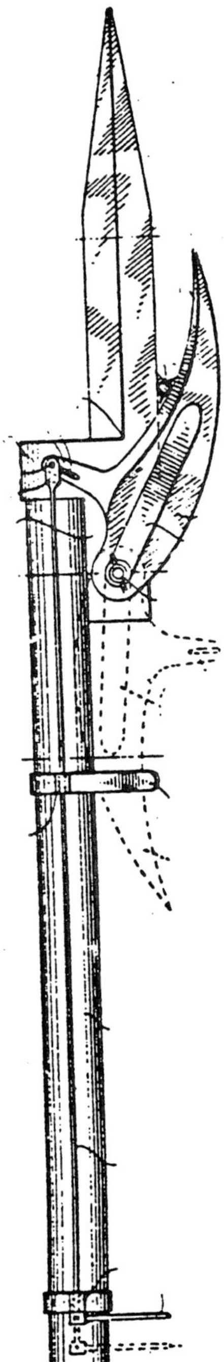

PAT.#1,273,287
Patented July 23, 1918 by I. Stankewicz
A combination bayonet/wire cutter/shear of gre strength yet is easily detachable and economical manufacture.

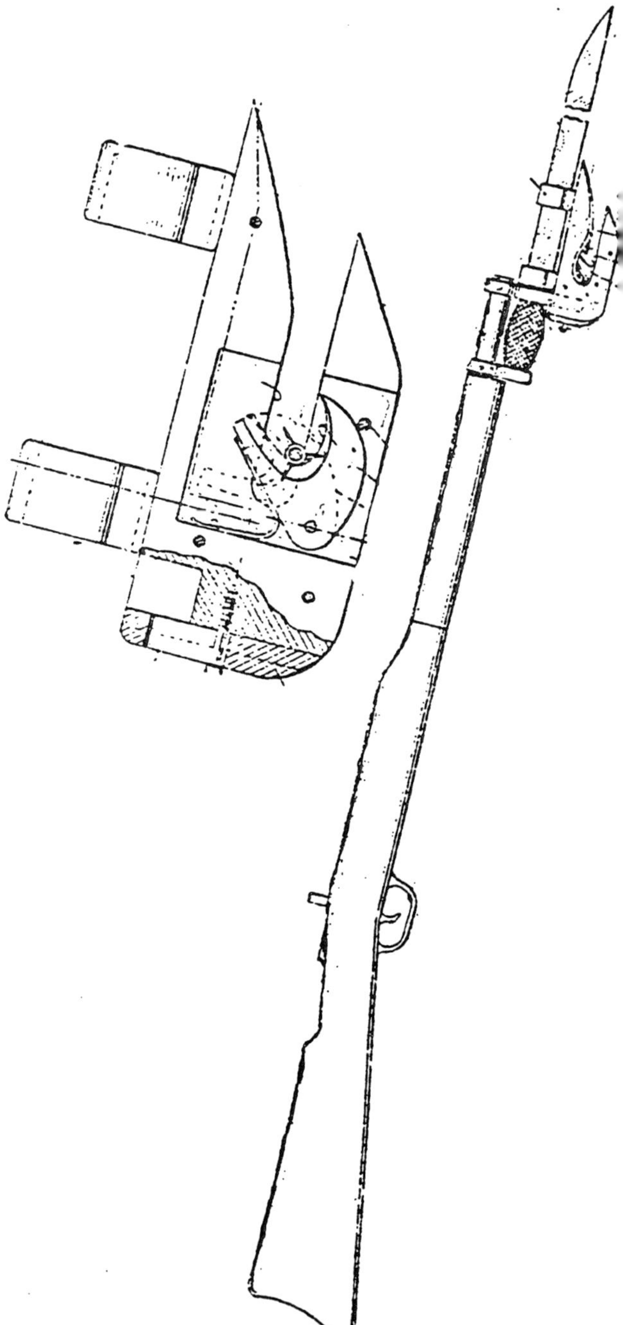

PAT.#1,298,166
Patented Mar. 25, 1919 by P. Borszcz
**A combination bayonet/wire cutter which can be used easily both mounted on a rifle or detached and held in hands.**

PAT.#1,297,571
Patented Mar.18, 1919 by M. Kaczmaryk
**A combination bayonet/wire cutter with design to prevent lateral movement of wire while cutting.**

FIG. 1.

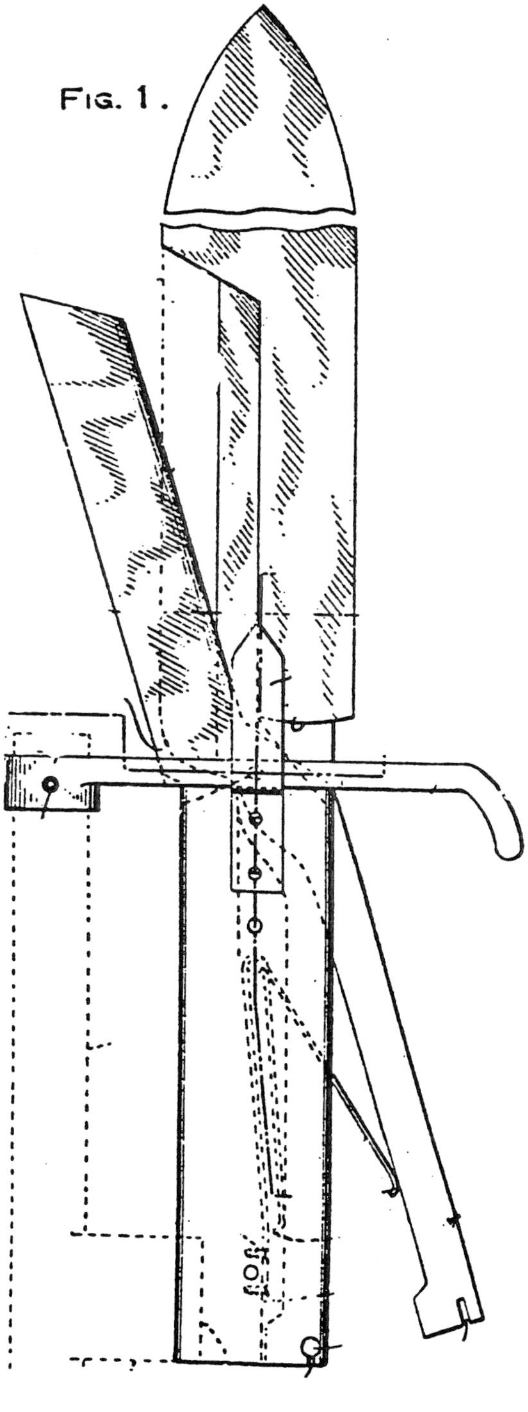

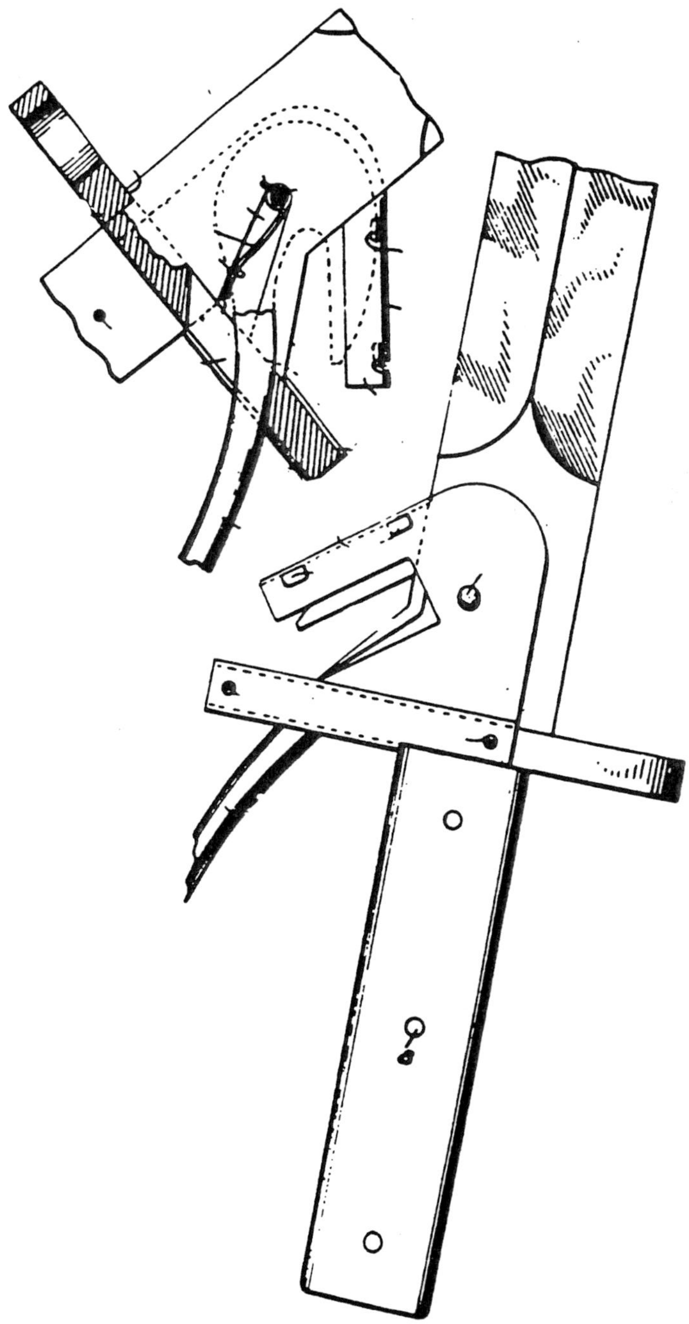

**PAT.#1,289,159**
**Patented Dec. 31, 1918 by J. Gramza**
**A combination bayonet/wire cutter, spring loaded to open automatically upon thrusting forward.**

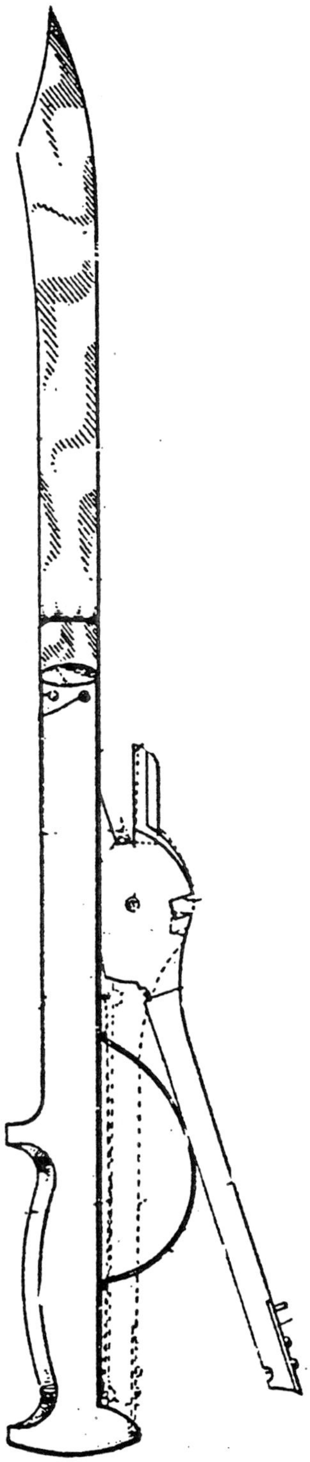

**PAT.#1,287,569**
**Patented Dec.10, 1918 by J. Bizon**
**A combination bayonet/wire cutter offering goo service without unsightly appearance.**

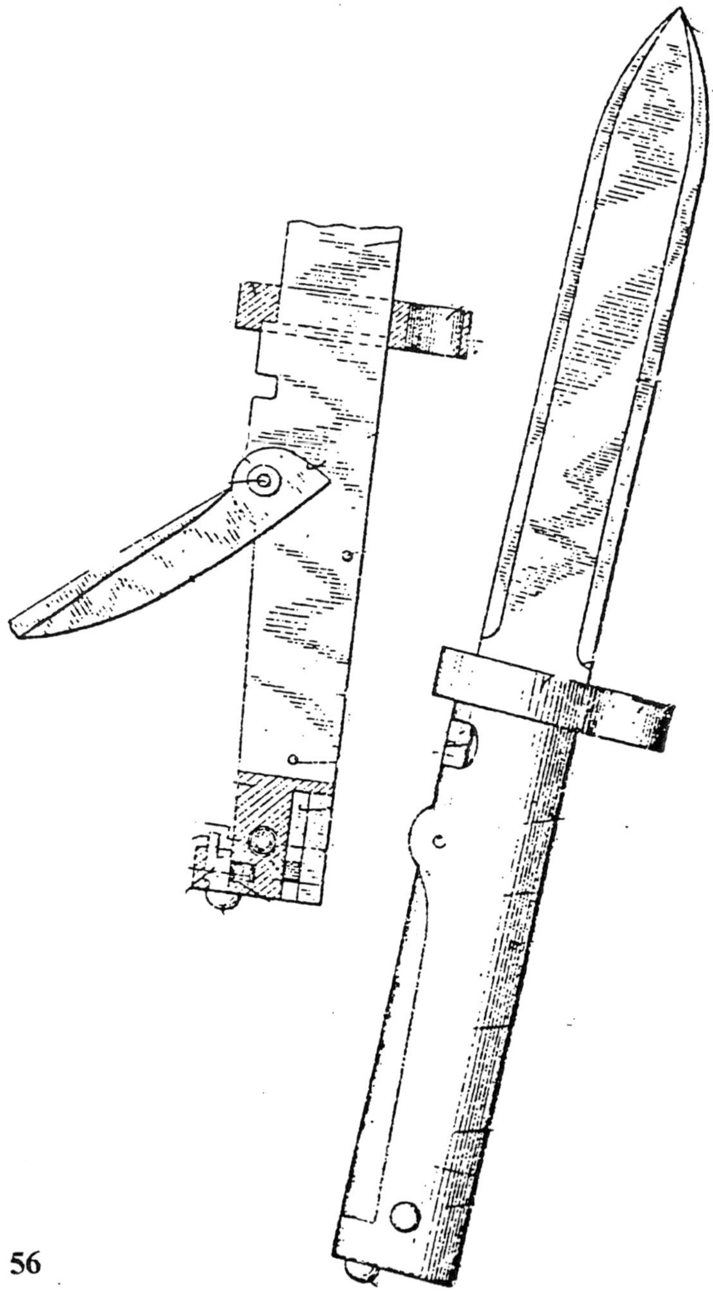

**PAT.# 1,303,277**

**Patented May 13, 1919 by J. Federkiewicz and M. Olszak.**

**A combination bayonet/ wire cutter improvement leaving no unsightly appearance yet is readily available for use.**

**PAT.#1,299,279**

**Patented April 1, 1919 by S. Wytrwal**

**A combination bayonet/wire cutter that opens like a folding knife.**

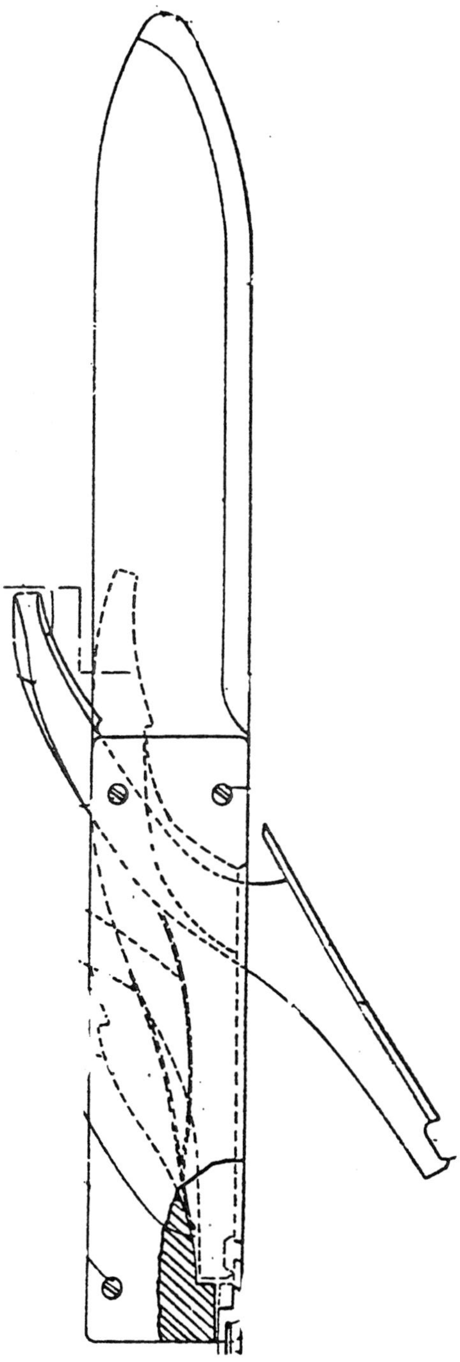

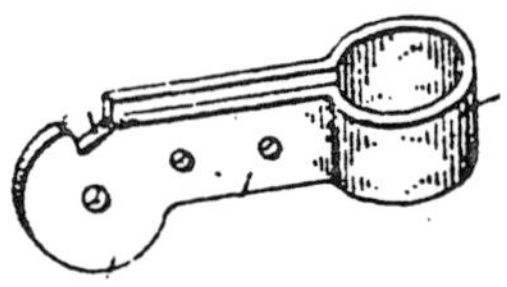

FIG. 6.

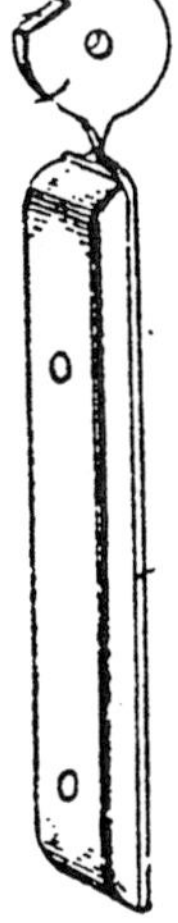

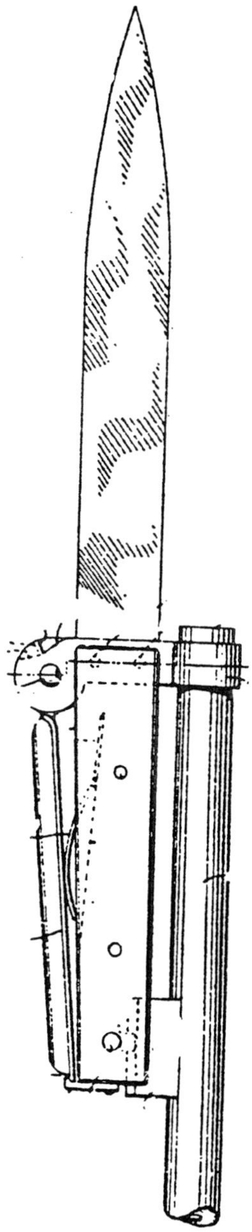

**PAT.#1,301,753**
**Patented April 22, 1919 by S. Seniw**
**A combination bayonet/wire cutter/shears with finger loops that fold out of the way.**

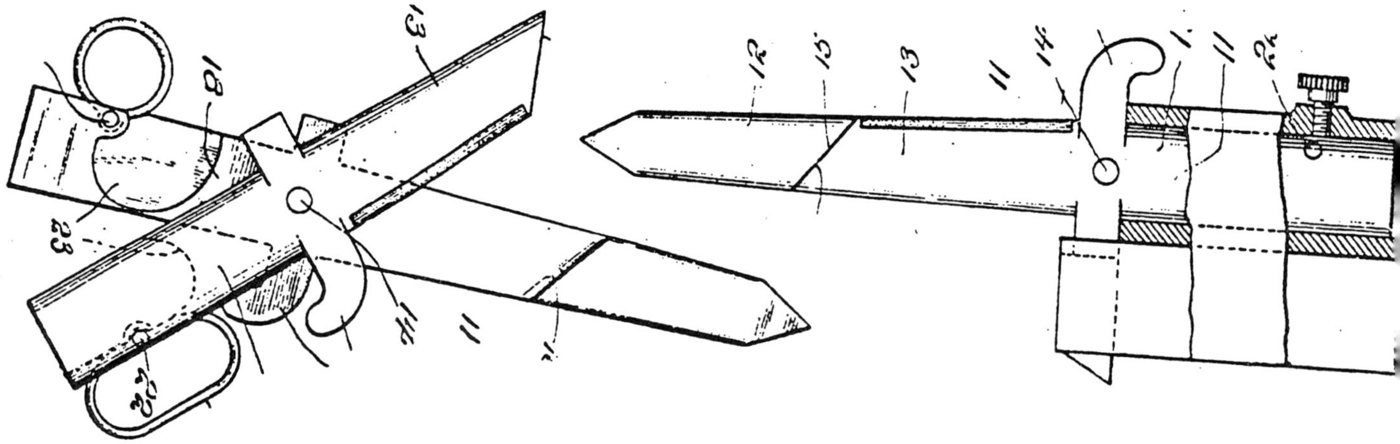

**PAT.#1,316,664**
**Patented Sept. 23, 1919 by W. Zadorosny**
**A combination bayonet/wire cutter attachable to any conventional model rifle.**

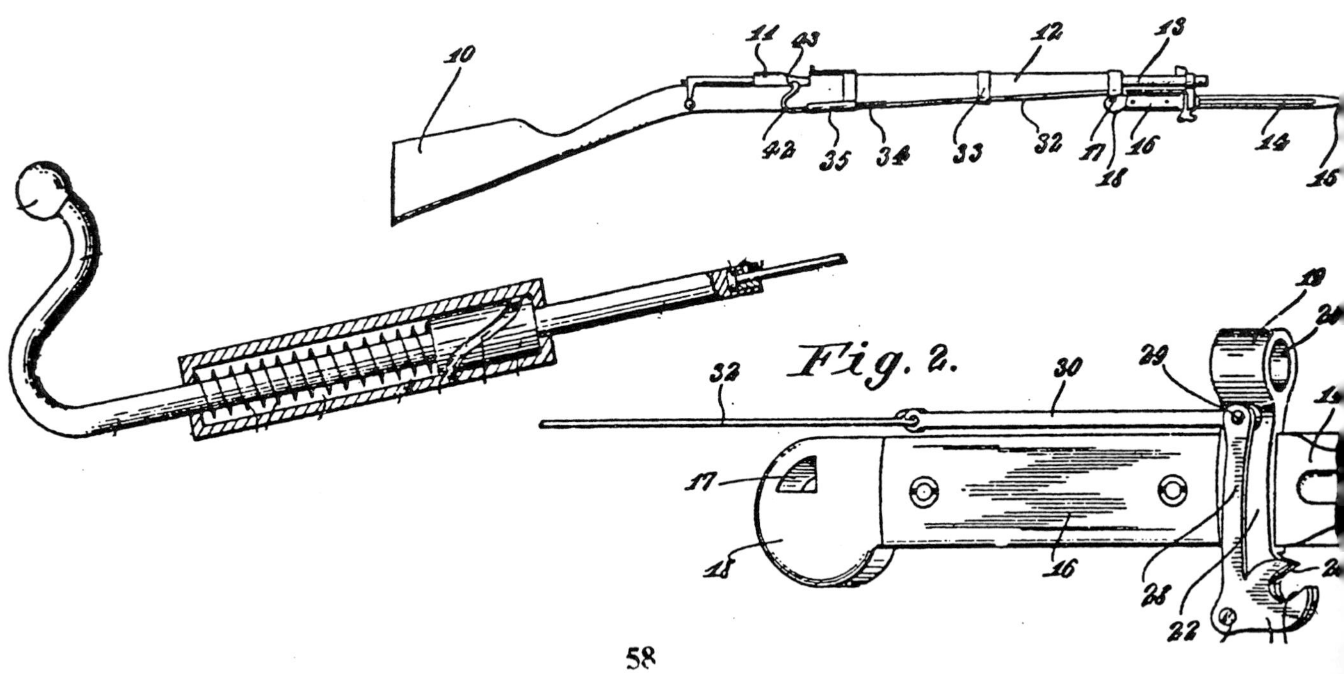

**PAT.#1,298,449**
**Patented Mar. 25, 1919 by G. Broadmack**
**A combination bayonet/wire cutter actuated by a pull cable, allowing operator to stay in safe position.**

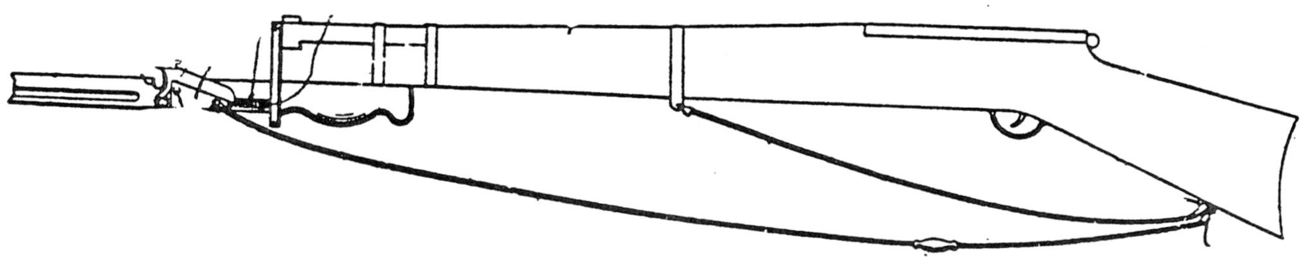

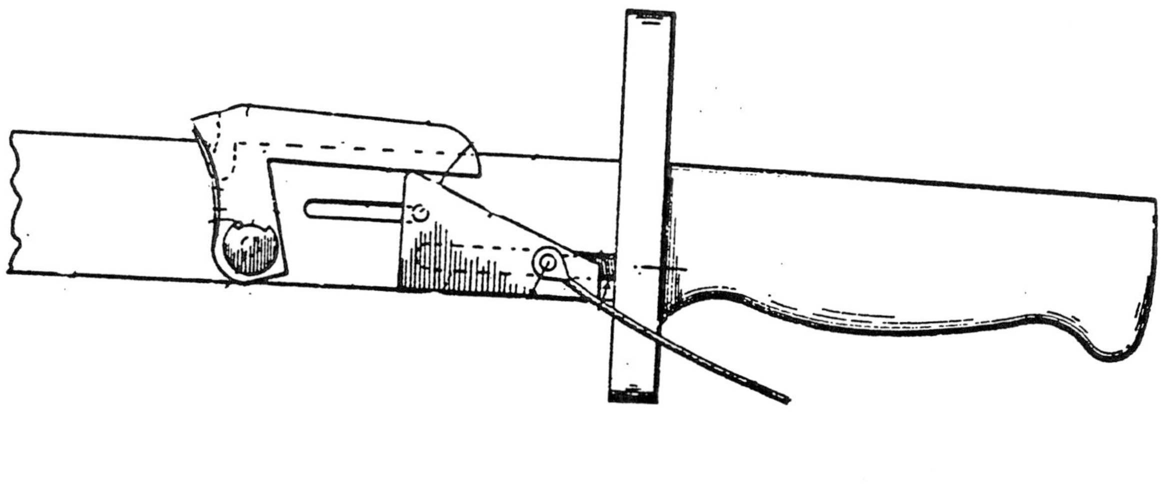

**PAT. # 1,342,382**

**Patented June 1, 1920 by John Hmirak**

**A combination bayonet/wire cutter utilizing serrated teeth and cutter rollers.**

**PAT.#1,317,632**

**Patented Sept. 30, 1919 by M. Krasnyanik**

**A combination bayonet/ wire cutter attachme**

**allowing wires to be severed at a distance.**

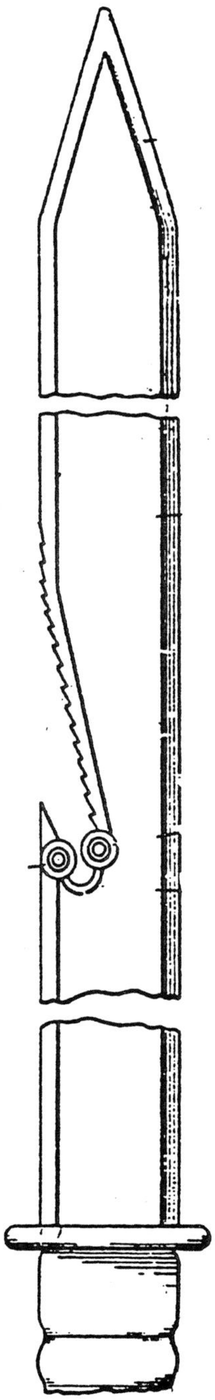

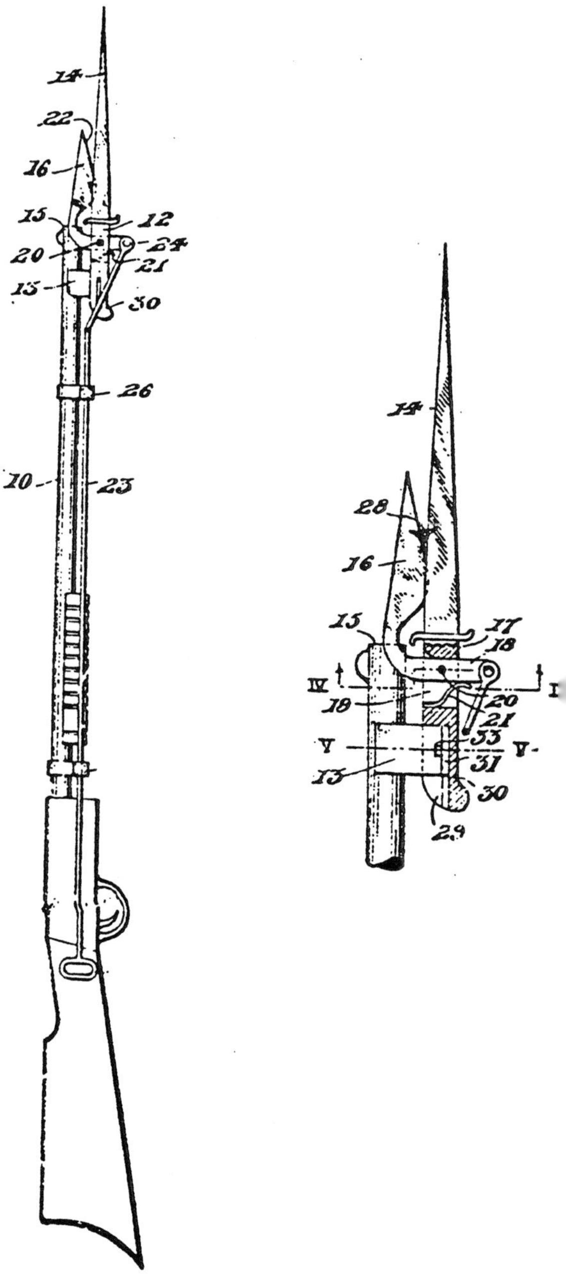

**PAT.#1,298,403**
**Patented Mar. 25, 1919 by H. Romanoff**
**A wire cutter attachment for rifles and other means. Easily attached and economical to build.**

**PAT.#1,300,940**
**Patented Apr. 15, 1919 by J.A. Erikson**
**A wire cutter attachment for rifles allowing the operator to keep weapon in firing position .**

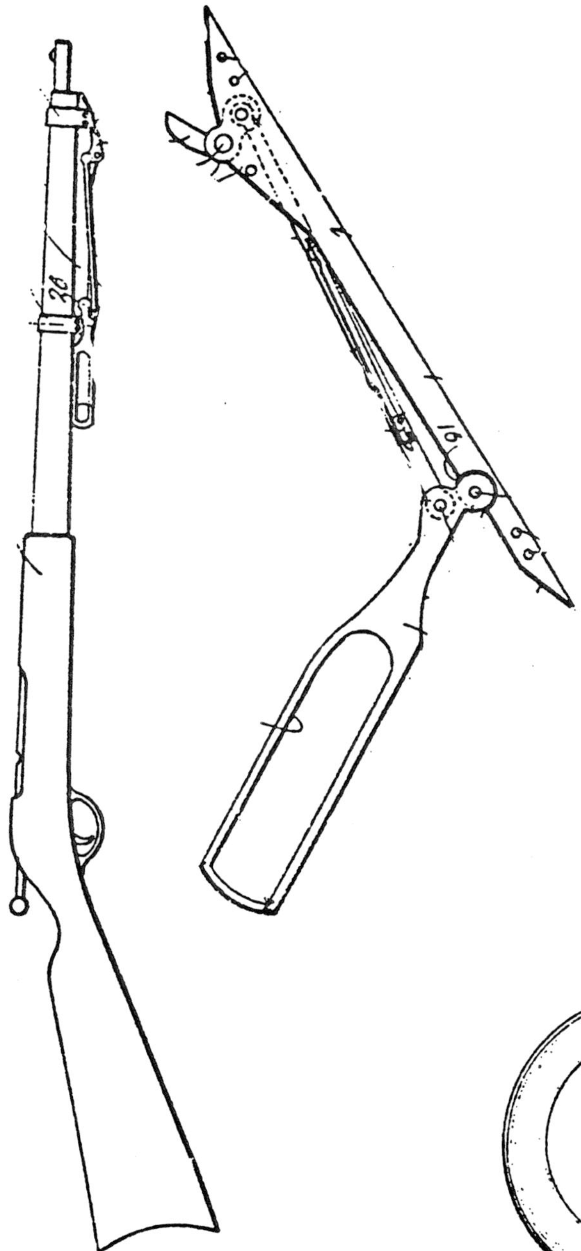

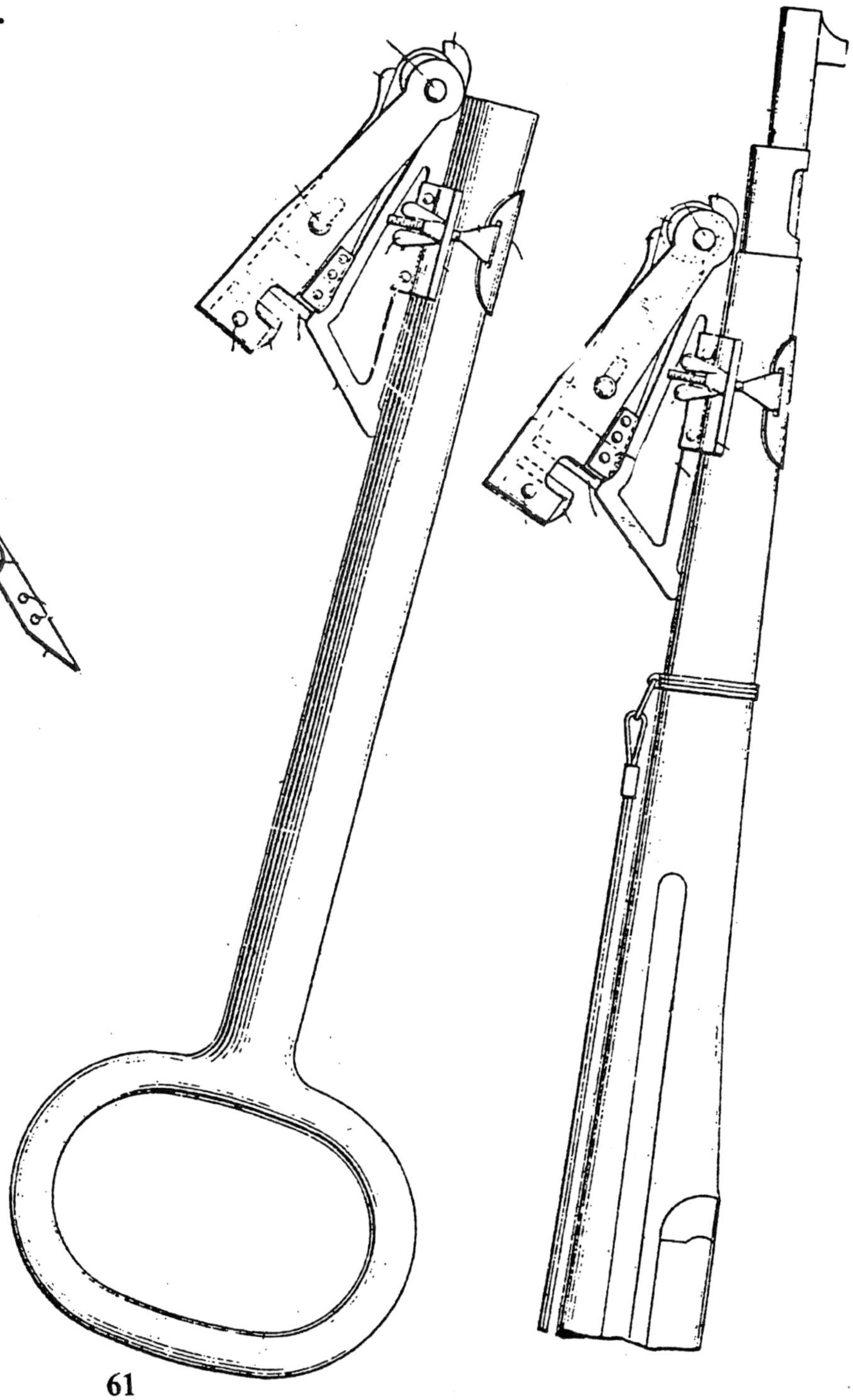

PAT.#1,147,631

Patented July 20, 1915 by G.D. James

A wire cutter easily attached to a rifle, and providing operator with obstacle clearing capacity.

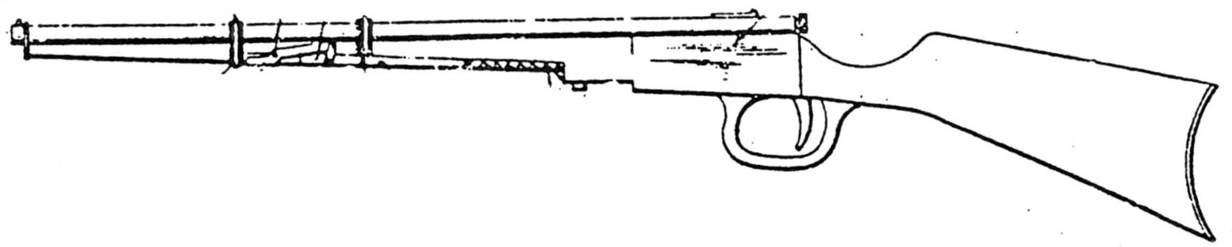

PAT.#1,333,556

Patented Mar. 9, 1920 by J.J.R. Meyer

A wire cutter attachment for rifles with revolving replacemant blades.

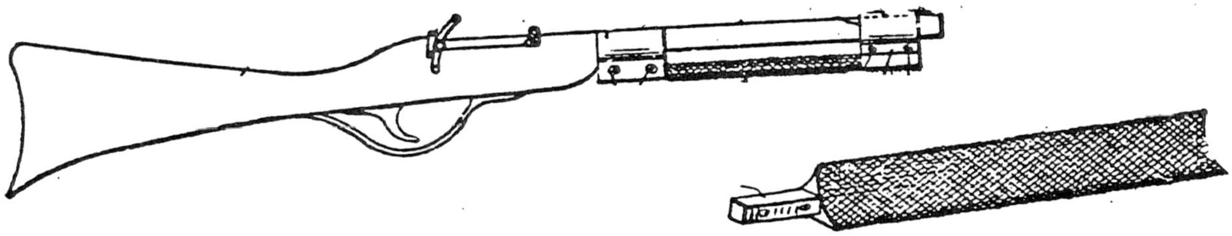

PAT.#1,241,455

Patented Sept. 25, 1917 by F.A. Warner

A wire cutter easily attached to all issue rifle stocks, providing operator wire cutting ability.

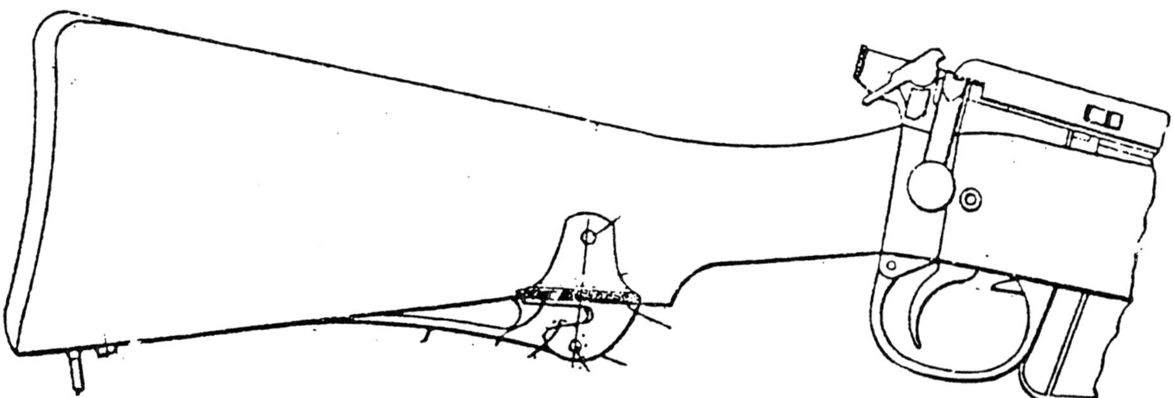

**PAT.#3,465,743**
**Patented Sept. 9, 1969 by W.J. Gilmore**
**A flat metalic tape, transported on reels, and being punched and twisted into spirals as it is unreeled on site.**

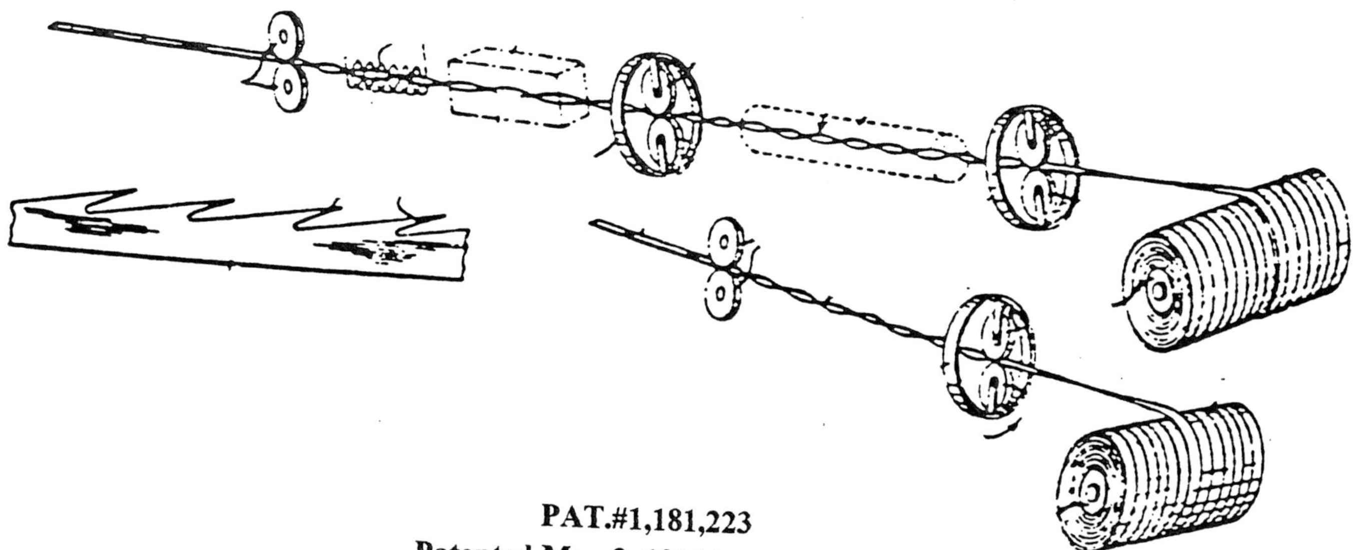

**PAT.#1,181,223**
**Patented May 2, 1916 by M. Holm**
**A shooter-cutter attachment for small arms, easily attached on bayonet. Operator pushes rifle forward engaging wire, then fires weapon. The projectile engages the elevated lever pushing it downward thus severing the wire. Advantages include quick action, from a safe position, and the wooden stock provides insulation from electrical shock. Operator is always a safe distance from the entanglement wire.**

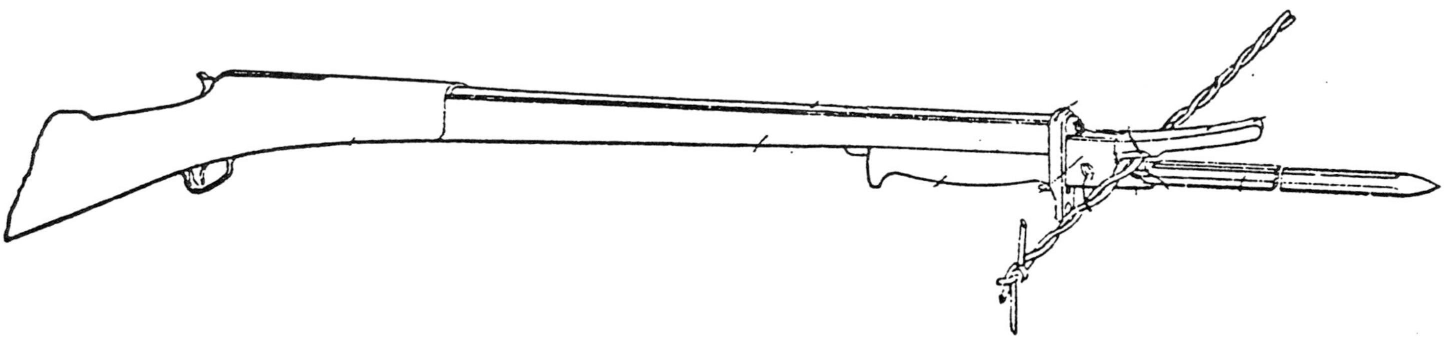

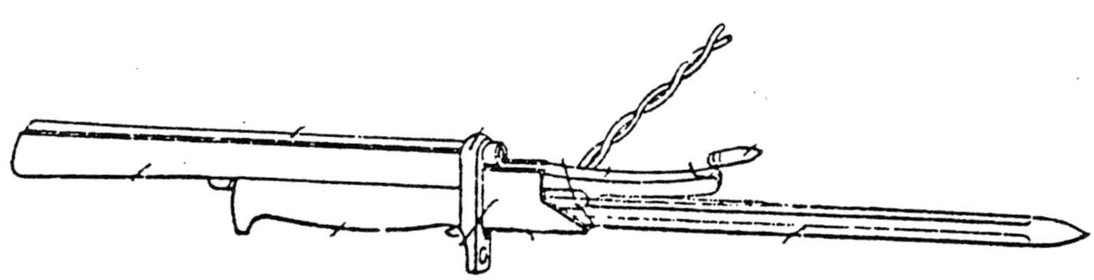

**PAT.#1,285,681**
**Patented Nov. 26, 1918 by C. Gora**
**A combination revolver/hand knife/wire cutter device for use in warfare.**

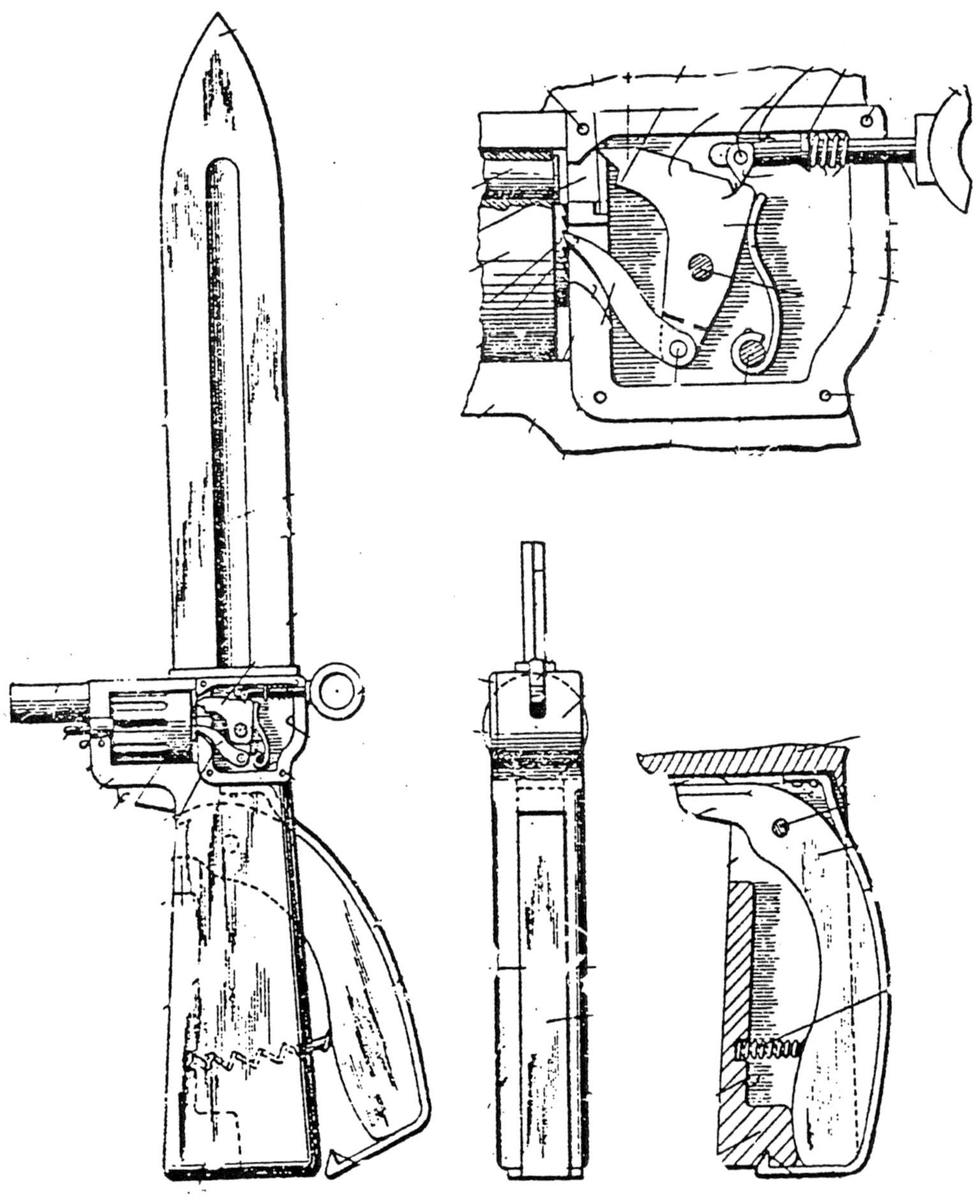

**PAT.#1,232,575**

**Patented July 10, 1917 by G. A. Lefort**

**A portable device, easily assembled and disassembled, used to take the place of regular obstacle fences.**

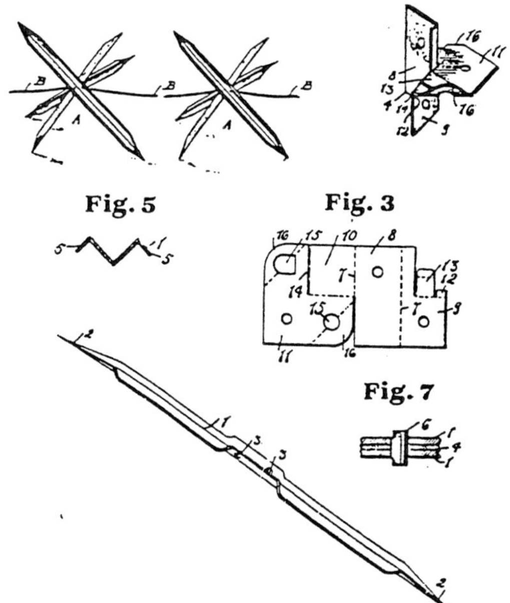


**PAT.#1,232,627**

**Patented July 10, 1917 by R. Vioch**

**A device designed to cut and drag obstacle wire. It is to be slipped over the muzzle of a cannon, aimed at the obstruction and fired. The projectile will receive the device and carry it through the obstacle, clearing the way for passage.**

**PAT.#3,463,455**

**Patented Aug. 26, 1969 by Paul Meckel**

**Design shows a canister housing a helical coil of metal tape with barbs. Canister is a part of a barbed wire patent.**

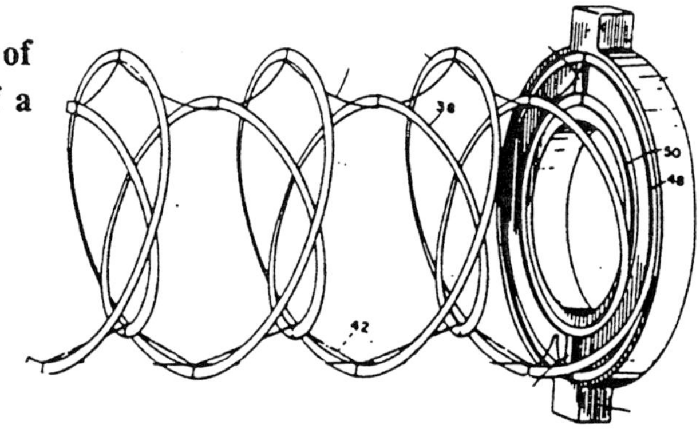

**PAT.#1,211,779**

**Patented Jan. 9, 1917 by J.A. Steinmetz**

**A shooter-cutter attachment for all small arms, easily attached by thumb screw. Push device to wire and shoot. The projectile will sever wire.**

Fig. 2.

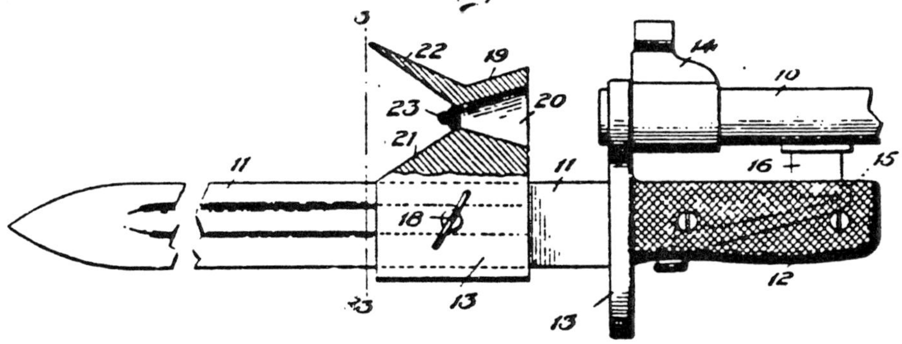


Fig. 4.

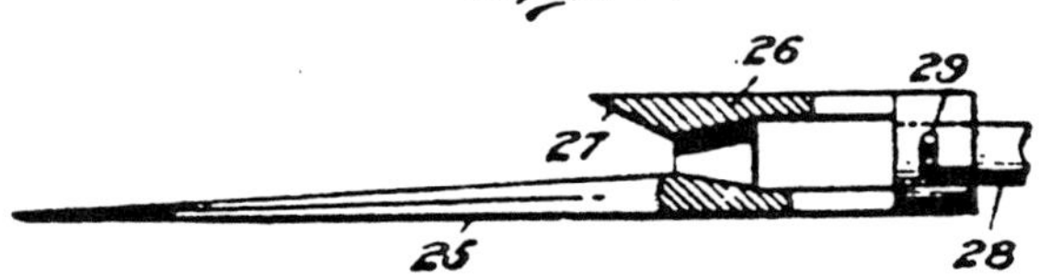


Fig. 5.

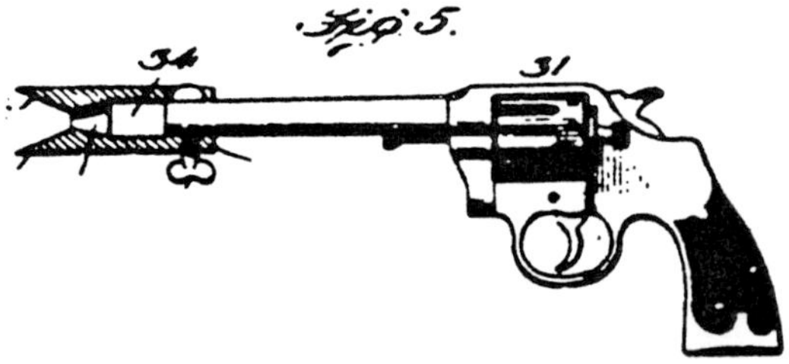

**PAT.#1,181,724**

**Patented May 2, 1916 by J.E. Barrows**

**A hand operated wire cutter designed to easily sever all obstacle wire and all parts are replaceable including blades.**

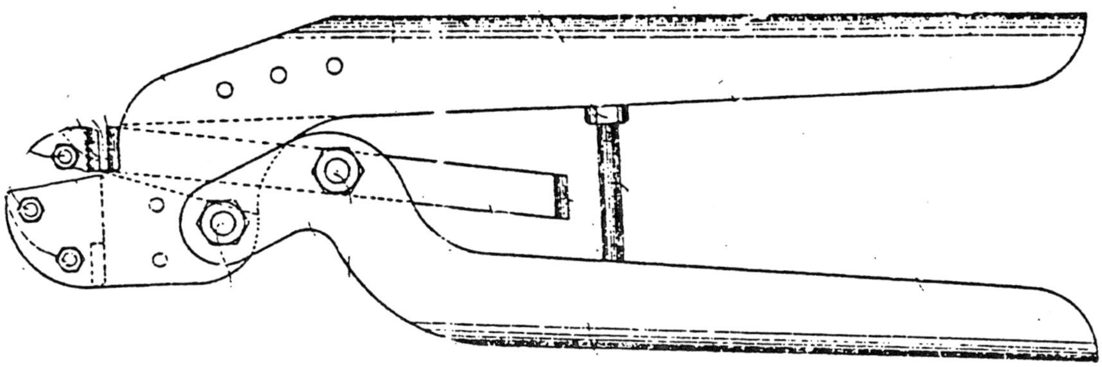

**PAT.#1,360,620**

**Patented Nov. 30, 1920 by J.G. Carew-Gibson**

**A heavy wire cutter/metal shear for all military and non-military uses.**

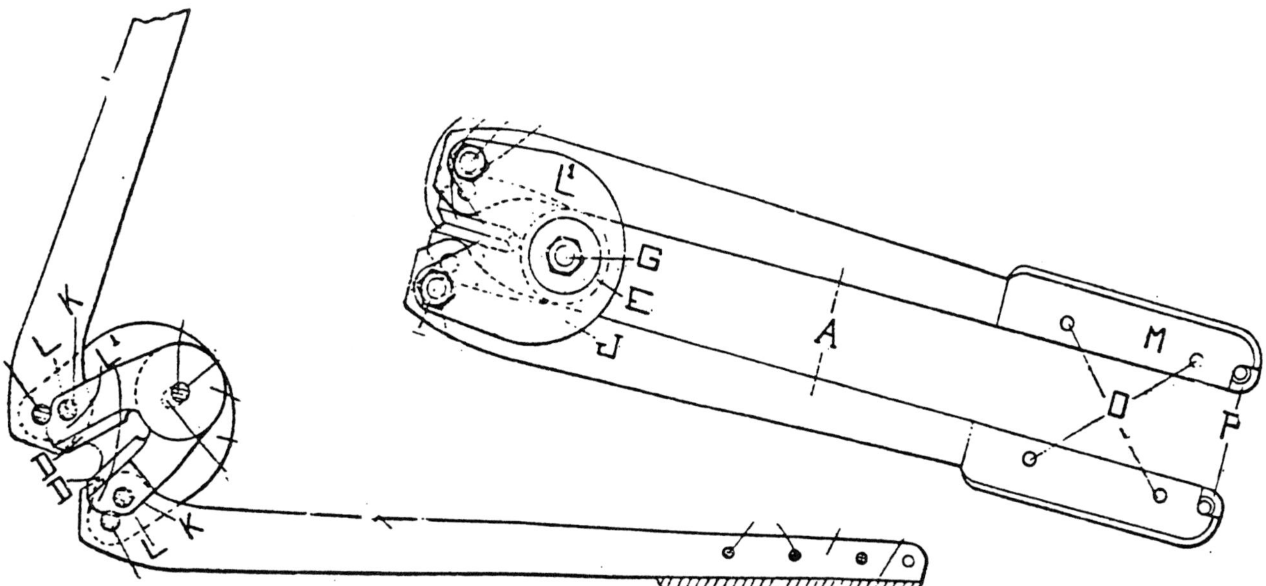

**PAT.#919,1132**
**Patented Apr. 20, 1909 by E.T. Conley**
**A combination pick, mattock, shovel, spade, wire cutter, and tent pole extention.**

**PAT.#1,164,765**
**Patented Dec. 21, 1915 by A.W. Slee**
**An attachment for regular issue hatchet to use in cutting entanglement wire.**

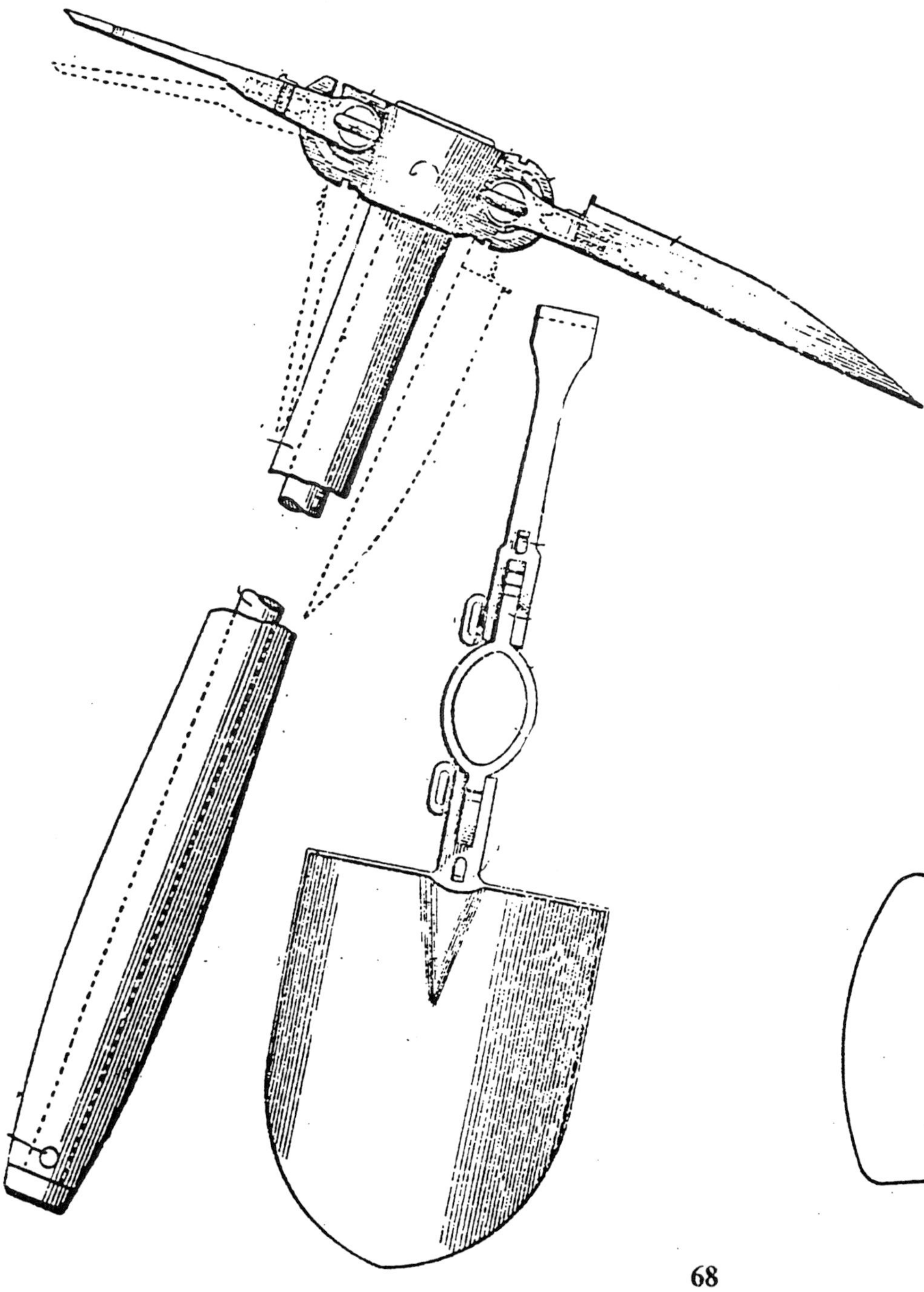

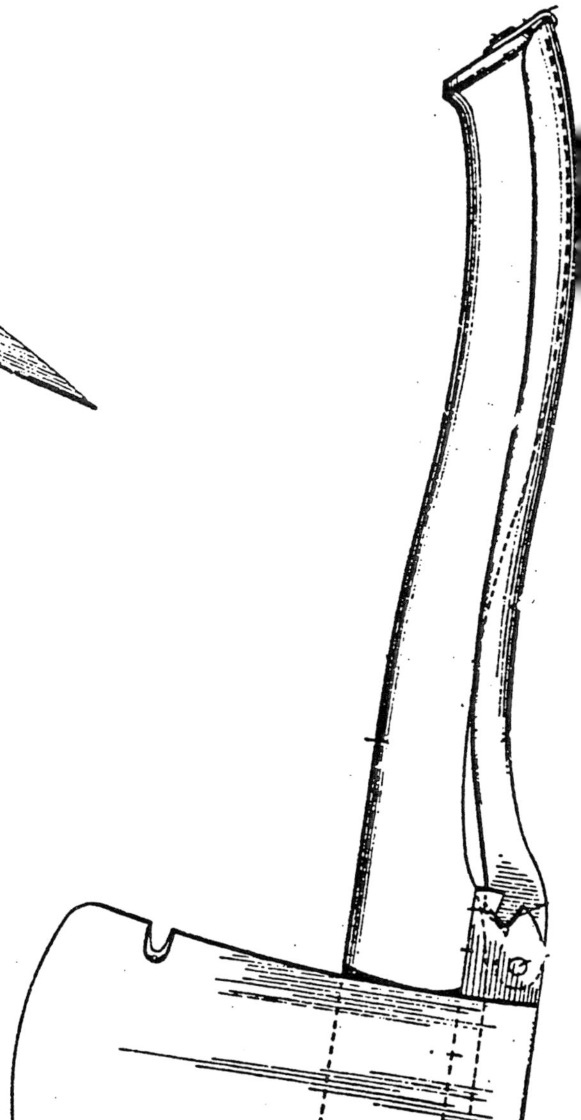

PAT.#1,252,537
**Patented Jan. 8, 1918 by A. Bergner**
**A combination bayonet wire cutter utilizing a thrust forward action to actuate the cutter.**

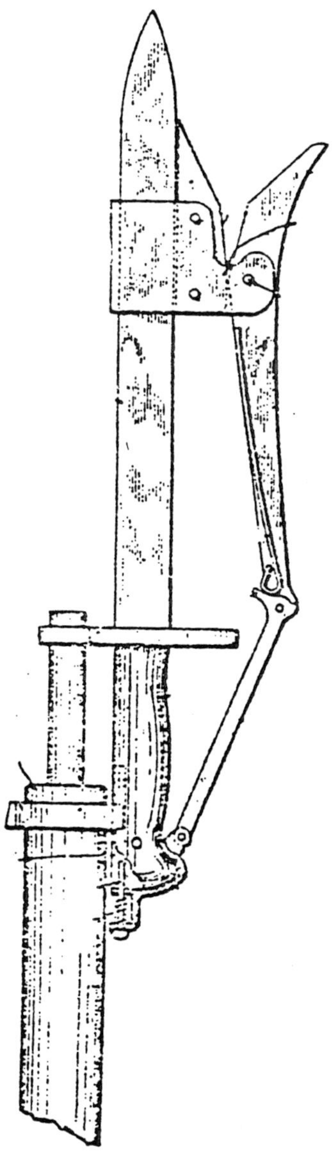

PAT.#1,265,151
**Patented May 7, 1918 by E. Wilczek**
**A combination bayonet/wire cutter utilizing rollers and blades to sever wire.**

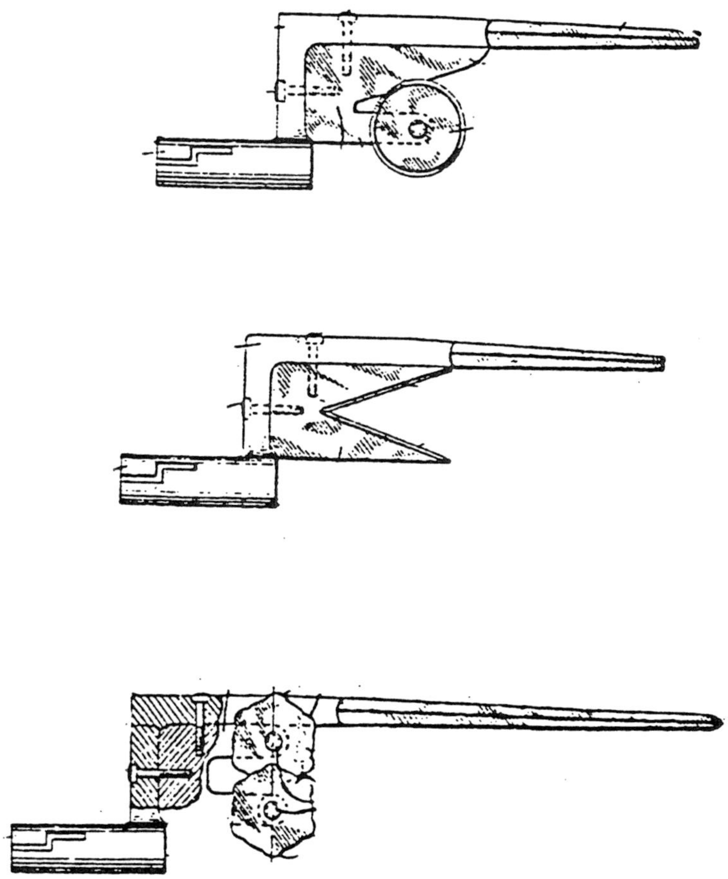

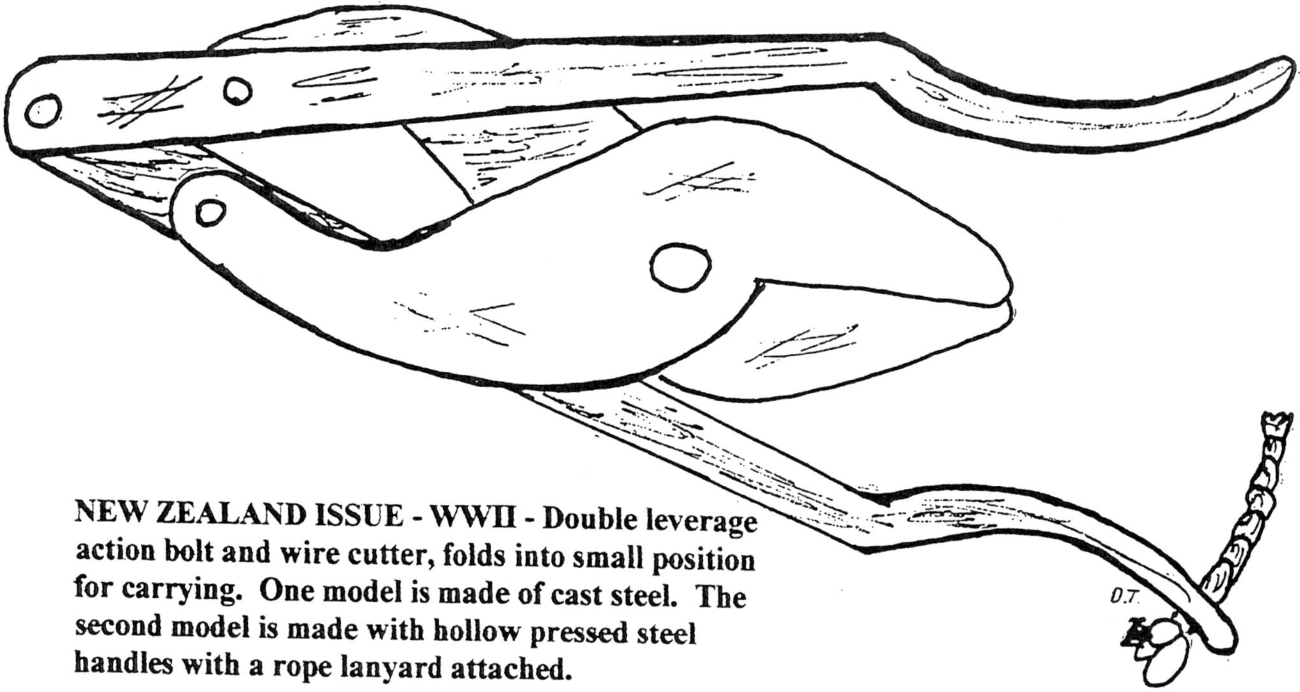


**NEW ZEALAND ISSUE - WWII** - Double leverage action bolt and wire cutter, folds into small position for carrying. One model is made of cast steel. The second model is made with hollow pressed steel handles with a rope lanyard attached.

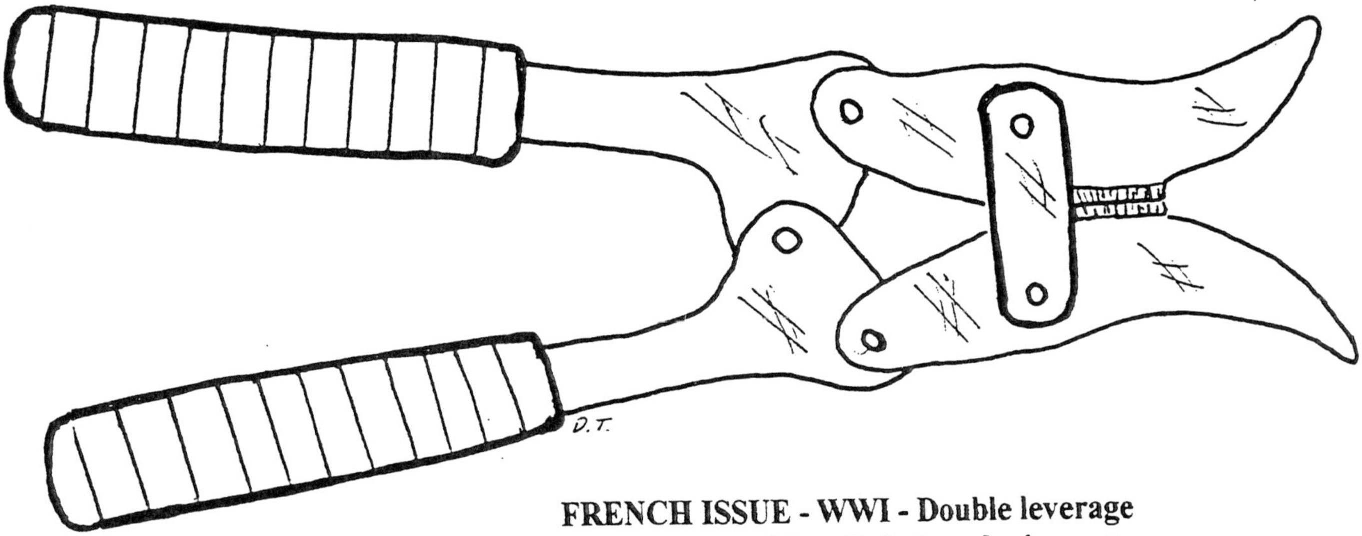


**FRENCH ISSUE - WWI** - Double leverage action, insulated handle bolt and wire cutter.

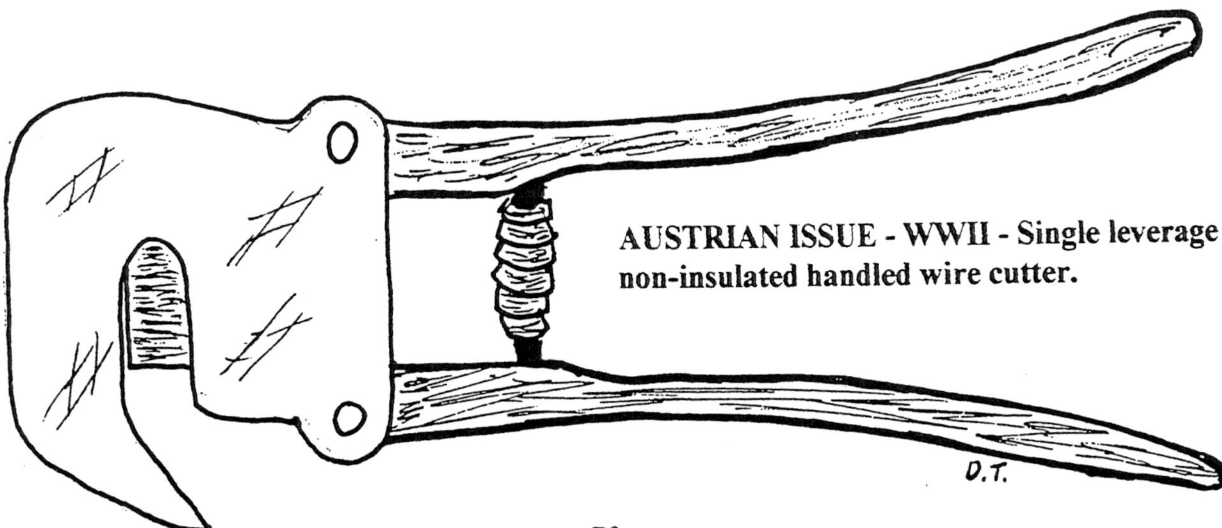


**AUSTRIAN ISSUE - WWII** - Single leverage action non-insulated handled wire cutter.

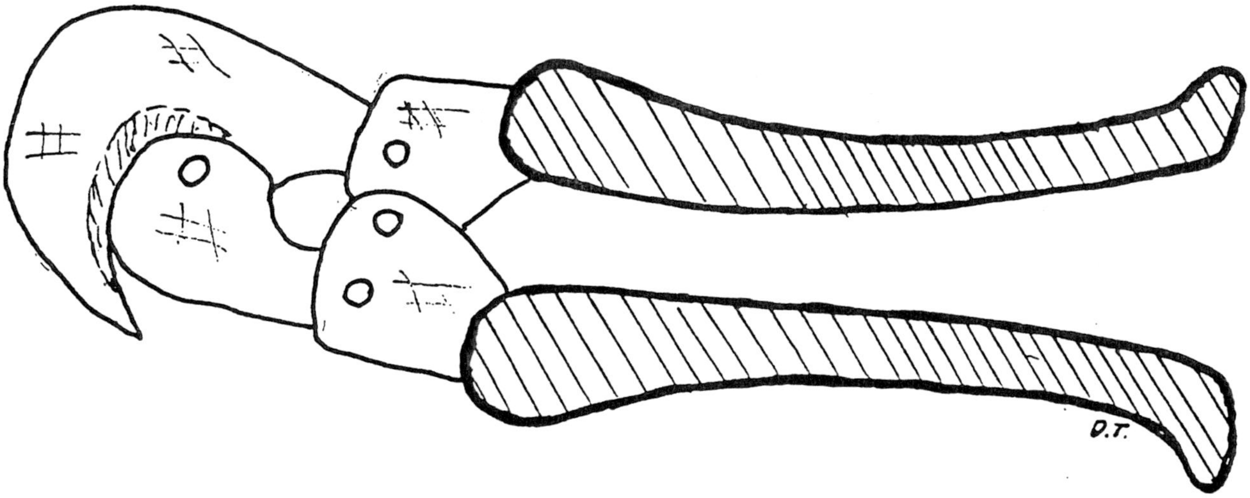


**ENGLISH ISSUE - WWII - Double leverage action with insulated handles for wire cutting.**

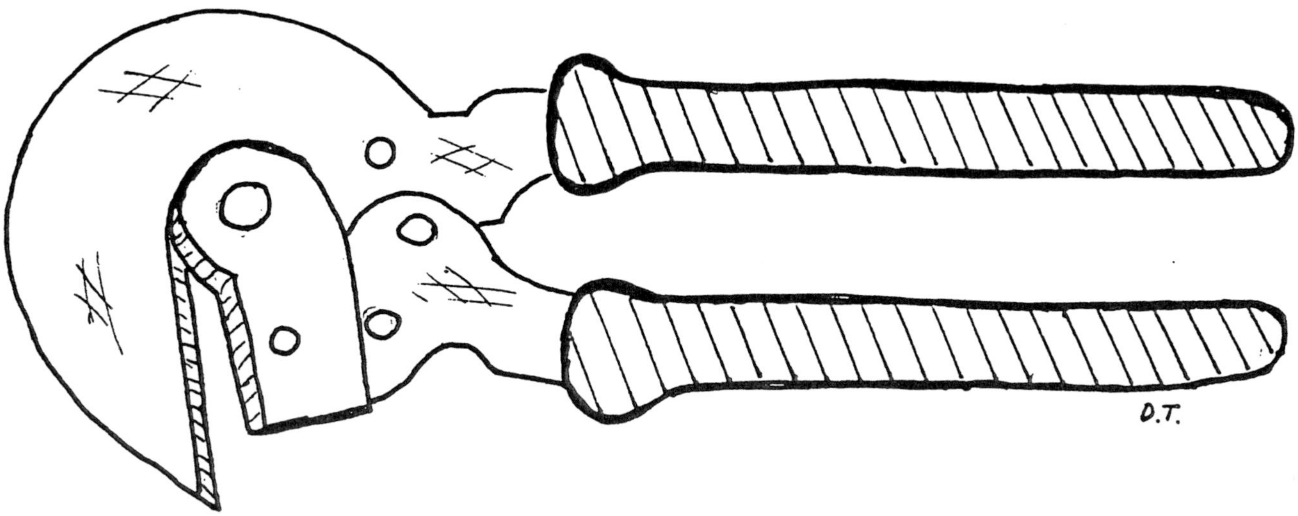


**U.S. MARINE ISSUE - WWII - Hook bill, double leverage action, insulated handled wire cutter.**

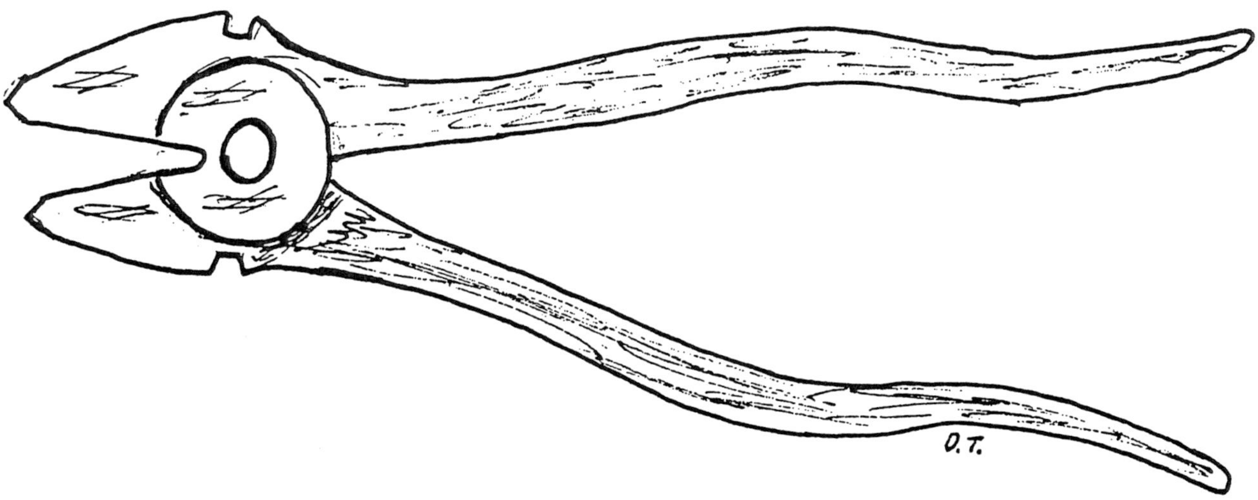


**U.S. ARMY ISSUE - WWI - Drop forged, single leverage action wire cutter. Designed for light weight easy use.**

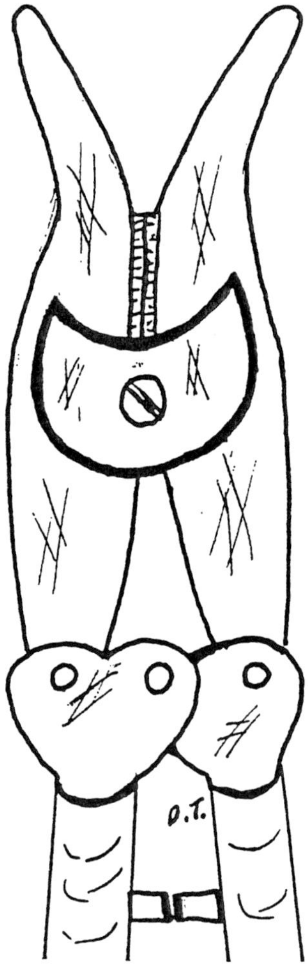


GERMAN ISSUE - WWII - Double leverage action with long handles, for bolt, cable, and wire cutting.

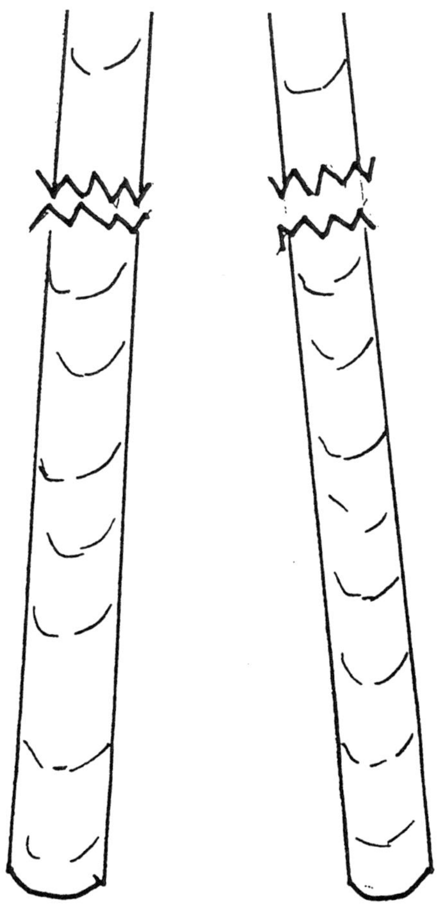

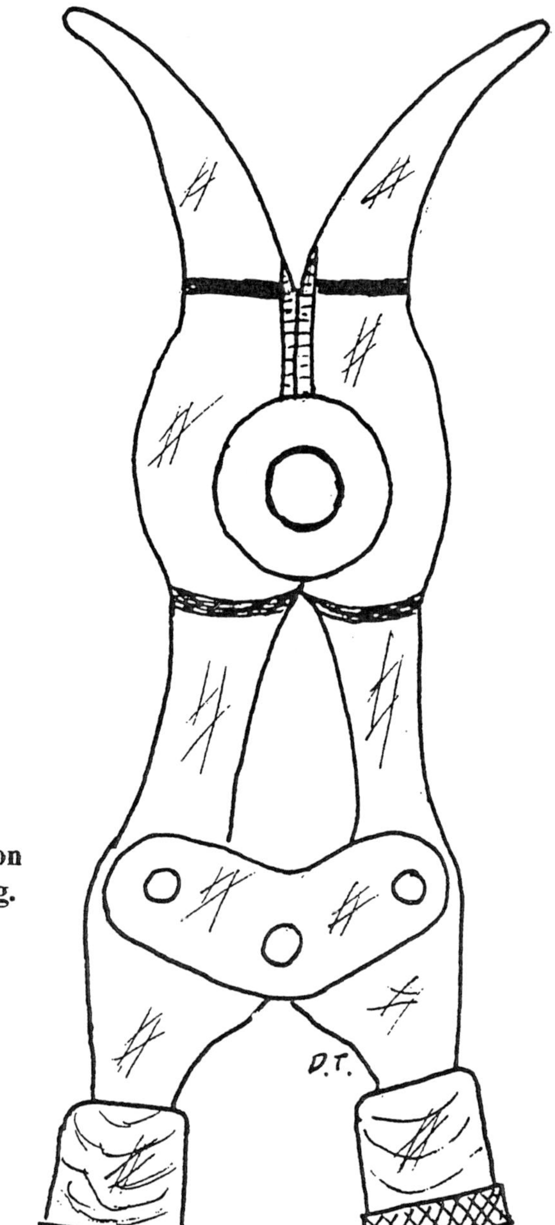


JAPANESE ISSUE -WWII - double leverage actio wire and bolt cutter with wooden handles for both weight reduction and insulating qualities.

A VOLUNTEER "PIONEER"

**Resembling a Knight of the Rountable in King Arthur days, this soldier is equipped to crawl forward to the barbed wire barriers and attempt to cut a path through.**

**Clad in heavy leather jacket and trousers, reinforced with chain mail mask and gloves, and armed with a long handled wire cutter, he pushed his iron shield along the ground in front until he reached the wire. Circa-WWI**

STOCKPILED WIRE SPOOLS WWI

BARRIER CONSTRUCTION UNDERWAY WWI

P. AMEDEO-ITALY

TYPICAL WIRE CUTTING EXPEDITION WWI

**BEHIND THE SHIELD**

One soldier guided and pushed the shield while the other places a tube of explosive underneath the wire barrier. WWI

**CUTTING THE BARBED WIRE**

It took a brave soldier to crawl to the wire and make a path for those behind.

P. AMEDEO-ITALY

**A DIFFICULT TASK AT BEST**

This rare photo shows the real story of trying to cut a path through the wire.
WWI

## SUGGESTIONS FOR FURTHER STUDY

Books- Barriers by Campbell & Allison

The Fencin' Tool Bible by Bill Marquis

Bayonets by Jerry L. Janzen

Bannerman Catalogue Of Military Goods

Experts- Jerry L. Janzen Tulsa, Oklahoma (Bayonets)

Homer M. Brett Alexandria, Virginia ( Military gear)

Billie Thornton Aurora, Colorado (Wire tools)

Larry Greer Kansas City, Misouri (Patents)

Museums- Devil's Rope Museum McLean, Texas
( largest barbed wire historical museum & library in world)

Kansas Barbed Wire Museum LaCrosse, Kansas
( dedicated to barbed wire history- library)

Magazine- The Barbed Wire Collector McLean, Texas

Other- The U.S. Patent Office Washington, DC

U.S. Army Military History Institute Carlisle Barracks, Pensylvania

# SECTION III.

## WARWIRE SPECIFICATIONS

**Specifications of the Ordinance Department for manufacturing various entanglement wires.**

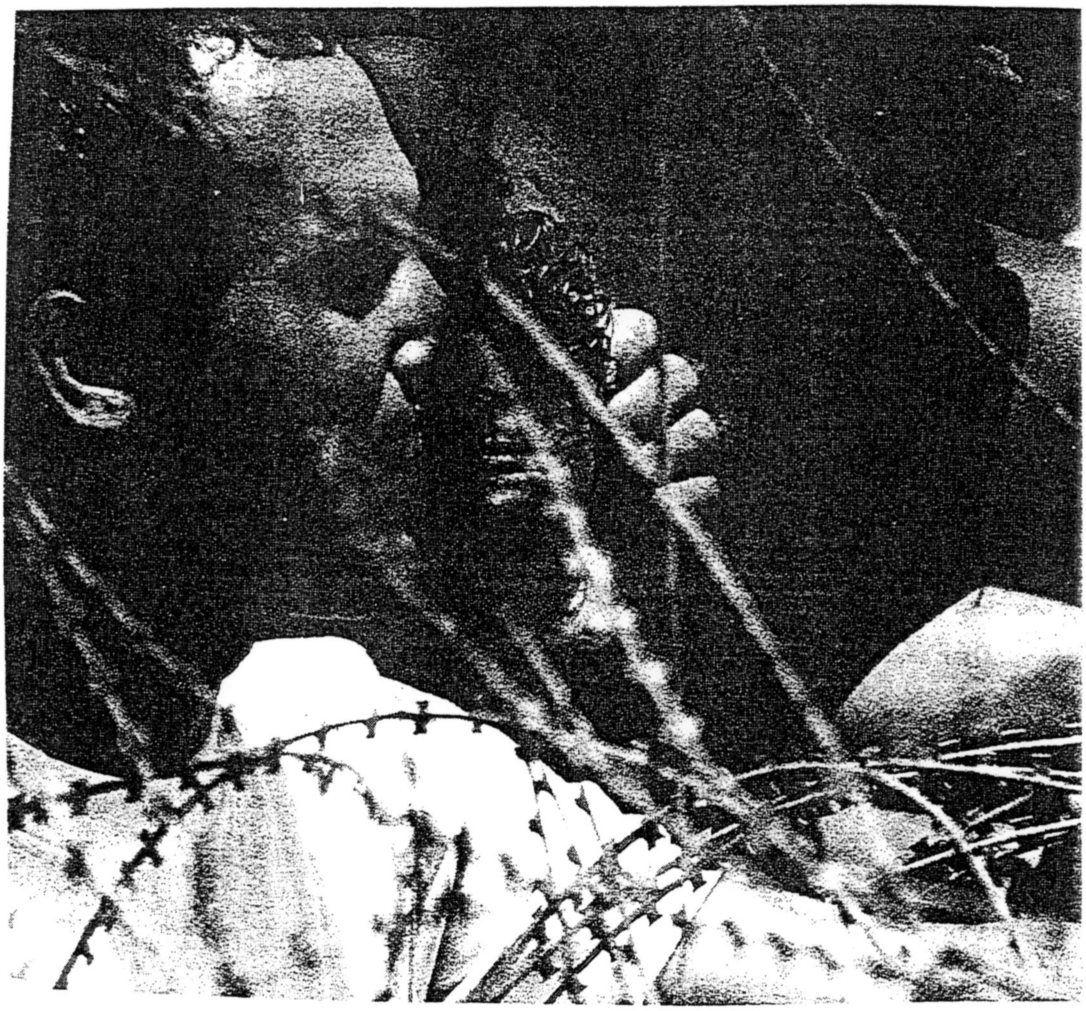

## WARWIRE SPECIFICATIONS

To date, research shows the year 1904 to be the first time U.S. military strategists issued "special specifications" for the manufacture of entanglement wire.

U.S. Army Pamplet No. 460, entitled Special Specifications For Steel Wire, was issued on August 15, 1917 by the Ordnance Department U.S. Army. A copy of this pamplet was purchased from the U.S. Army Military History Institute at Carlisle Barracks, PA, for research purposes.

A reference note contained on the cover, states it is to be used in connection with Ordnance Pamplet No. 445. No. 445 must have been published earlier, but it has not been found to date.

Other references listed on the cover, state the information within was first issued on January 14, 1904. The pamplet was revised in 1907, 1908, 1909, 1910, 1914, and again with the 1917 issue in hand. With these revisions unknown, we must look to the 1917 issue for wire specifications.

Random excerpts from this pamplet follow;

1. The wire must be of square cross-section, one-eighth or one-tenth inch on a side, corners rounded with a radius of 0.01 inch. It must be smooth and free from checks, twists, flaws, and other defects.

2. The wire shall be free from rust, withstand a breaking stress of 200,000 lbs. P.S.I., pass torsion and chemical tests as set up by the U.S. Army Ordnance Department.

3. The wire must be delivered neatly and snuggly wound on wooden spools of sufficient strength to withstand shipment and use while the wire is being unwound. The windings shall be in neat concentric layers to avoid tortional or other serious strains. Spools shall have a hole 2 ½ inches in diameter in the center for mounting on an arbor for unwinding. The spool shall not be less than 15 inches in diameter. The net weight of the wire on each spool shall be 185 lbs. plus or minus 10%.

The manual further describes in detail several different tests the wire must pass before it is accepted. Also, many shipping instructions are included as well as penalties imposed for rejected wire.

## "CONCERTINA" WIRE SPECIFICATIONS

Before presenting specifications of Concertina wire, we should clarify the terms used. The word "concertina" is derived from the Concertina accordian, which flexes in and out by the pulling of the musician's hands. This same action is used in stretching out compacted concertina coils of entanglement wire.

The concertina coils are actually made up of a round, single line, hard drawn, spring steel wire with a tensile strength of 200,000 P.S.I. The Army title is listed as MIL-STD-130. It cannot be cut with ordinary wire cutters. Extremely vicious 4 point barbs, 3/4 inch long, are attached at 2 ½ to 8 inch intervals. A slight bend is incorporated into the wire allowing it to naturally coil into approximate 40 inch diameter coils.

As the single wire comes from the machine, it is wound onto a 40 inch diameter spool until sufficient footage is reached to make a finished concertina coil. At this time, the single line coils are paired and clipped together in a parallel manner with the end coils made into permanent circles. The metal clips are placed at 72 inch intervals.

Two wire handles are attached to each end coil for handling purposes. The numerous coils are then compressed into small bundles and secured with tie wires. Each bundle will stretch out to a length of 56 feet for entanglement installation. The bundles were packed in bales of 28, secured by wooden boards and steel strapping. The length of the bales was not to exceed 8 feet to facilitate shipping by truck and rail.

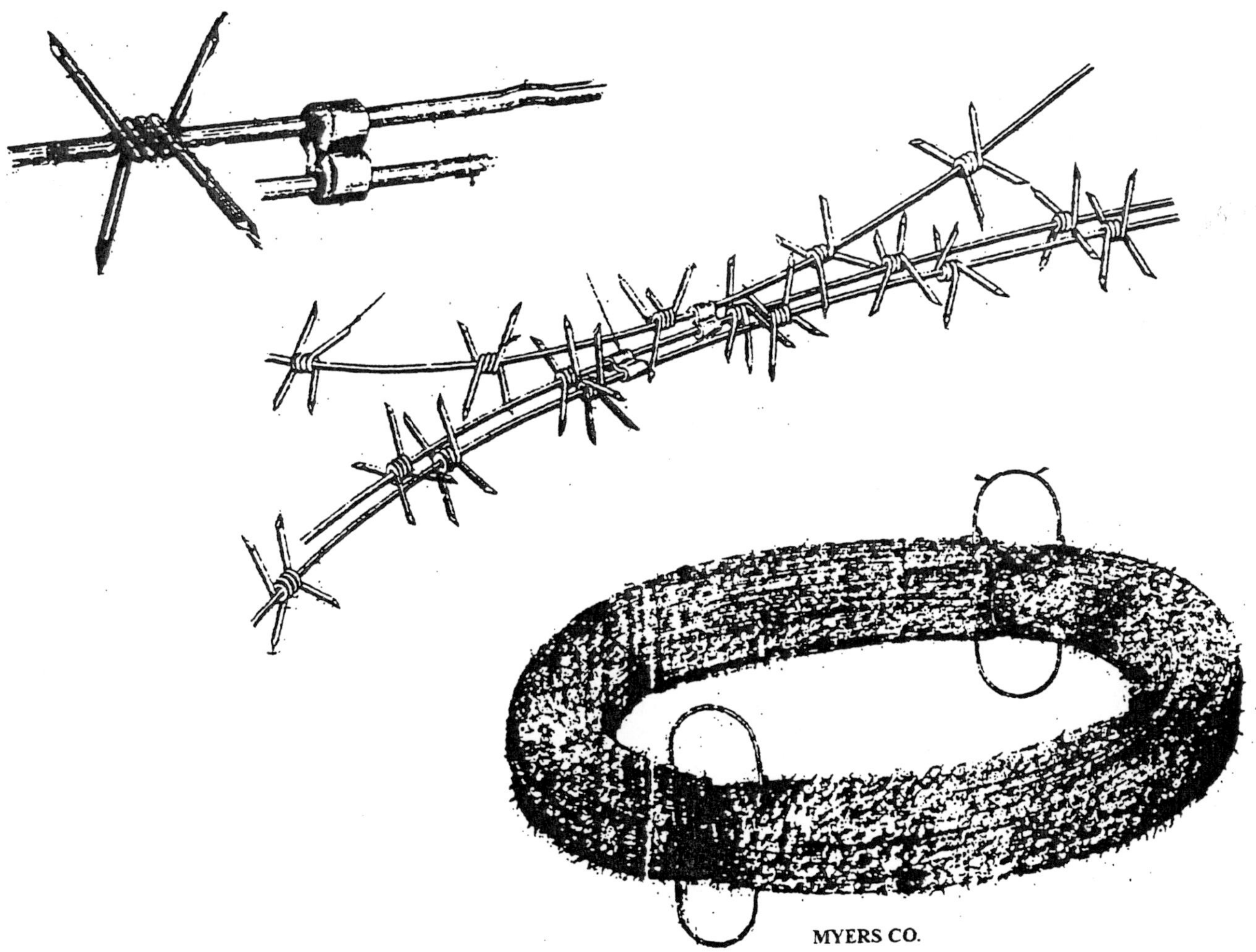

MYERS CO.

## SUGGESTIONS FOR FURTHER STUDY

Documents- Wire Entanglements Addenda No. 1 U. S. Army (1904-1918)
Military History Institute, Carlisle Barracks, Pensylvania

Manufacturing Contracts- Most U.S. Steel Companies, U.S. War Dept.

Layne Myers Manufacturing Company
( contract on file in Devil's Rope Museum Library)

# SECTION IV.

## BARRIER DESIGN AND INSTALLATION

**Taken from the U.S. Army Engineering Manual on how to construct various entanglement barriers.**

## INTRODUCTION TO SECTION IV.

## BARRIER DESIGN AND INSTALLATION

Barriers to hinder enemy advance have existed since the earliest beginnings of war. Among the first devices used was the "thorn caltrop" made from binding needle sharp thorns together and spreading them in the path of the charging aggressors.

Another early day barrier device was sharpened wooden stakes used in pits, moats and trenches. Sharpened stakes were also used atop earth works and set into walled fortifications. This use of stake barriers in front of defense positions could be called the fore-runner to barbed wire entanglements.

In 1670, during the Dutch War of Independence, the Dutch military found themselves weak in cavalry positions. To strengthen this weakness they devised a portable "trestle." This device was free-standing, made in various lengths, could easily be moved or chained together in lines of defense.

Its unique feature was in the front side facing the enemy where numerous wooden and metal spikes projected outward. Breeching such a barrier was formidable, and the Dutch prevailed over the enemy. Known as a "trestle horse," or Chevaux-de-Frise, the device has since taken on various forms and variations but has persisted until modern times.

Another variation of the Chevaux-de-Frise called the "knife rest," has also proved popular down through time. Lighter in weight, and simple in construction, the device uses barbed wire in diagonal bracing and barrier design. It provides easy-opening, protected gates for paths, roadways, streets and trenches.

There are many photographs dating from the Civil War to modern times where the Chevaux-de-Frise barriers are depicted at check-points of temporary stockades and other designated parameters. The advent of razor wire makes this ancient device even more successful in security control.

Until the turn of the century, most entanglement barriers were built by desperate men using anything at hand to protect themselves. This usually resulted in barrier construction of little pre-planning and of a type desired by whomever was in charge.

After 1900, military strategists throughout the world began to experiment and design barriers that could be constructed quickly, take advantage of terrain, and also protect the troops doing the work.

The role of entanglement barriers diminished as the pace of war quickened. Artillery bombardment and rapid deployment of war machines rendered entanglements obsolete. However, these same tactics increased the number of prisoners of war to unprecedented levels.

Confinement of these prisoners on the homefront among U.S. communities, required secure measures at best. Barbed wire used in various configurations and combinations provided this security successfully.

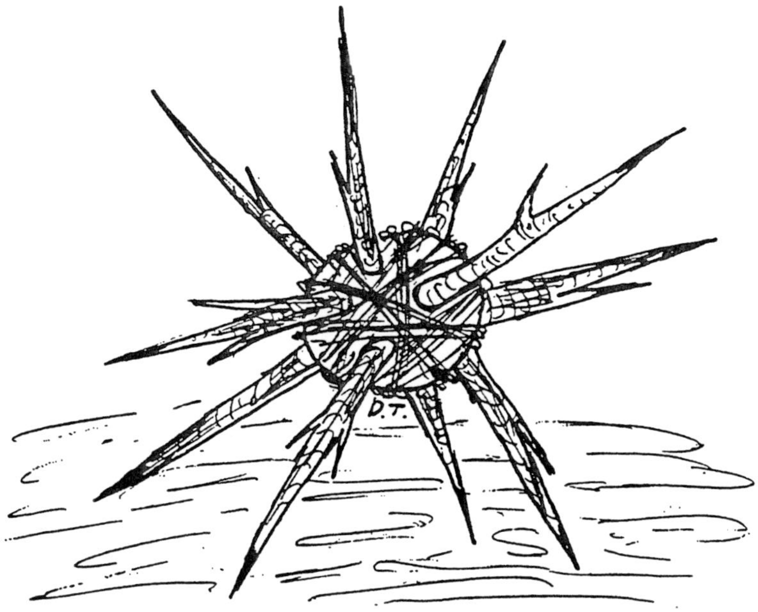

**THORN CALTROP**
Made by forming a ball of tree gum, clay, or mud, then thrusting thorns into the mass and wrapping it with sinew or strings and drying.

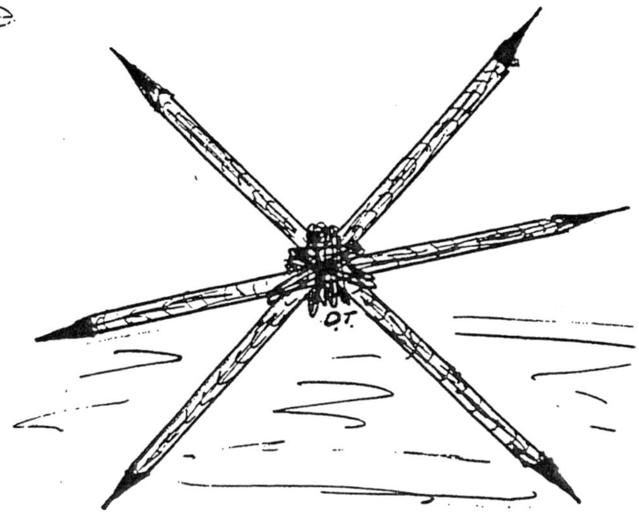

**STICK CALTROP**
Same as above only sharpened sticks are used instead of thorns.

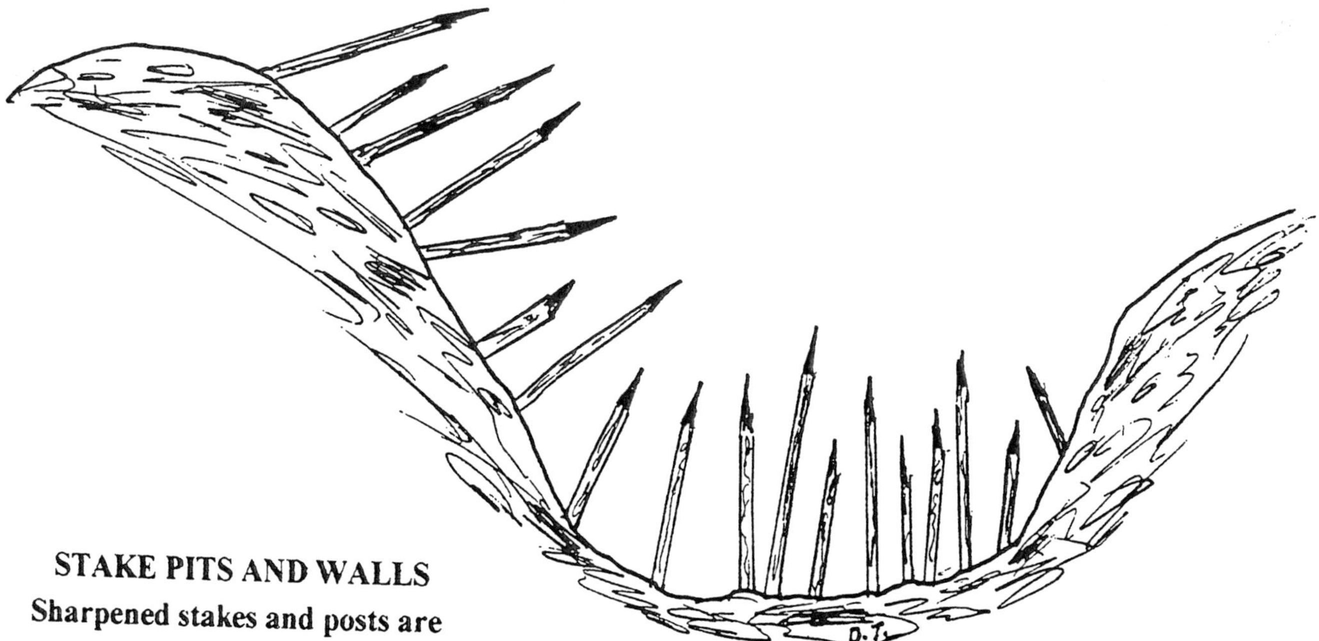

**STAKE PITS AND WALLS**
Sharpened stakes and posts are stationed in the bottoms of moats, ditches, and in walls or embankments to slow the enemy.

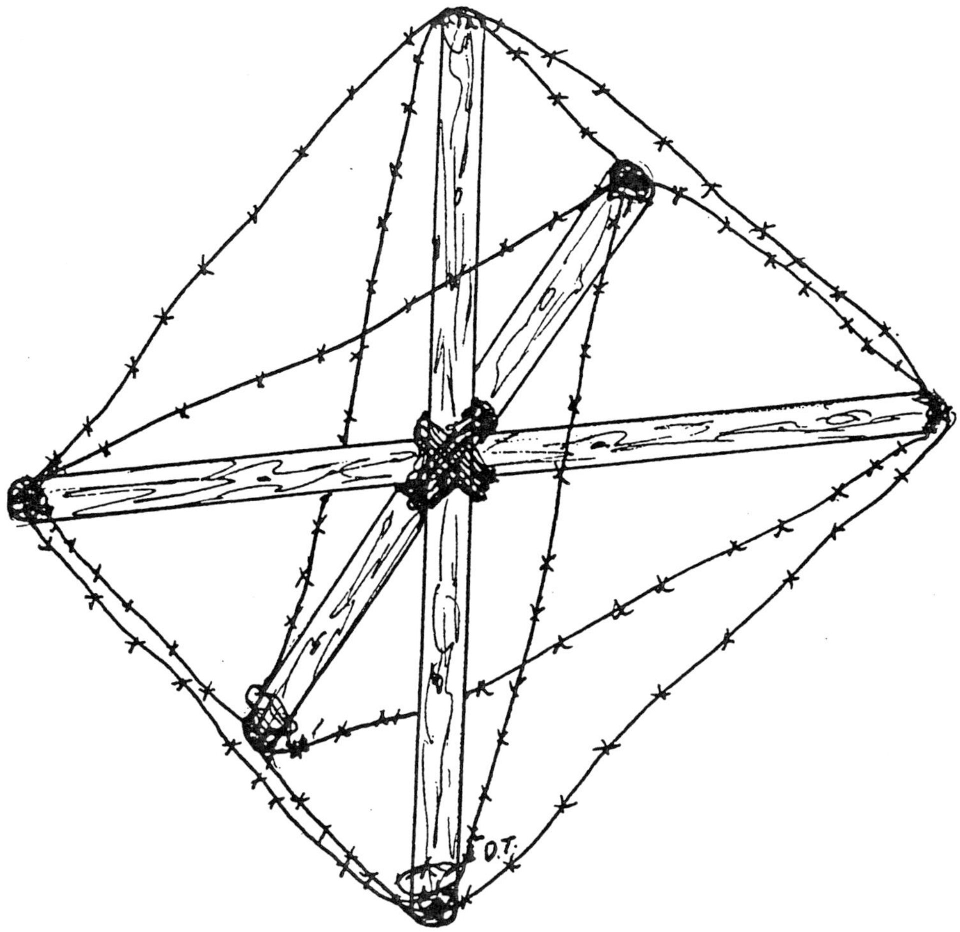


HEDGEHOG

A larger version of the caltrop, the early versions of the hedge hog were made from any materials at hand, bound with rawhide or rope and featured sharpened ends. They were made in many sizes and used to slow down the charge of oncoming troops.

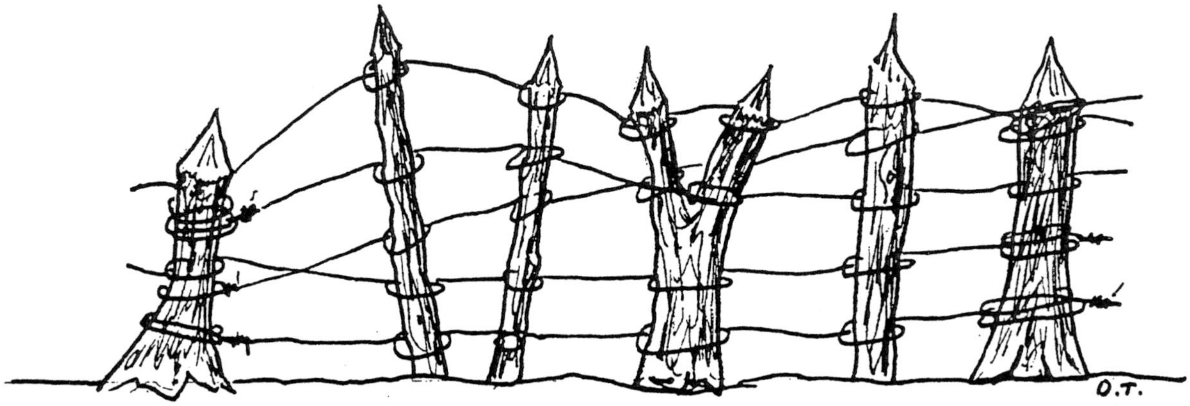


FIRST KNOWN ENTANGLEMENT DESIGN

Artist's conception of the telegraph wire entanglement used at Vicksburg. Based on a painting of the battle by a Civil War artist.

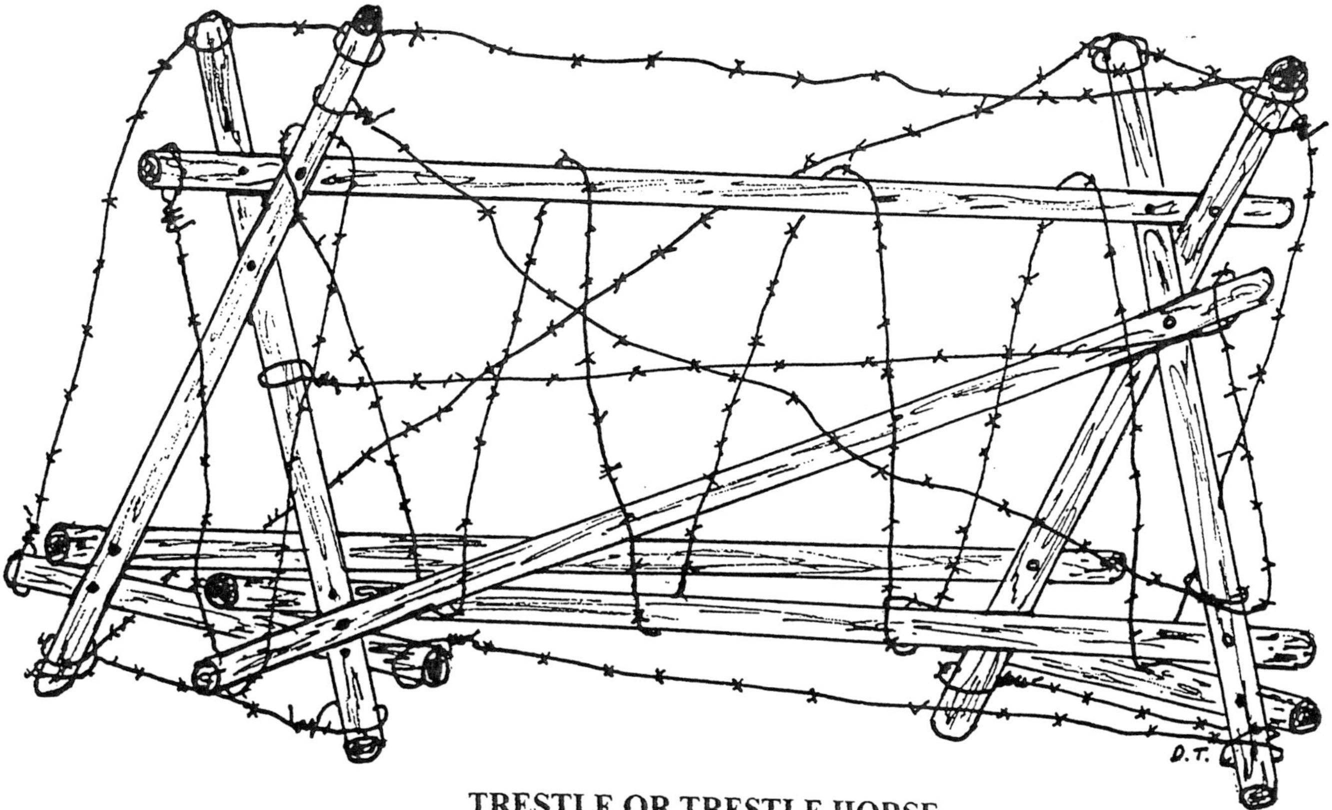


## TRESTLE OR TRESTLE HORSE

Earliest forms of the trestle were bound with rawhide or rope and used wooden spikes or stakes, thorns, or metal tipped spikes. The trestles became more effective after the advent of barbed wire and even more so in modern times with the use of razor wire.

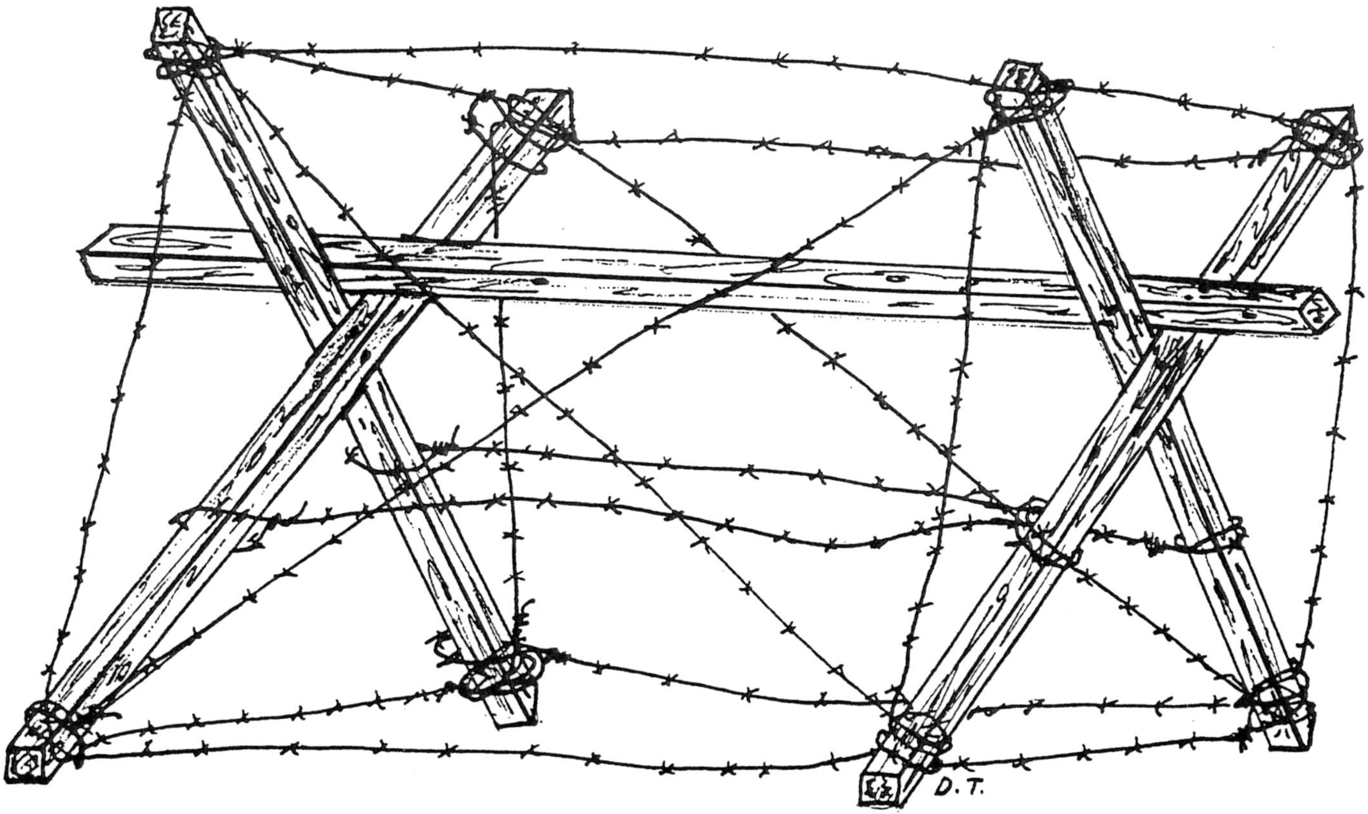


## KNIFE REST TRESTLE OR KNIFE REST HORSE

Earliest versions of the knife rest trestle used a spiked log in the center. These horses were chained or roped together in long lines to break up cavalry charges. The advent of wire made the bracing and stabilization of knife rest horses easier.

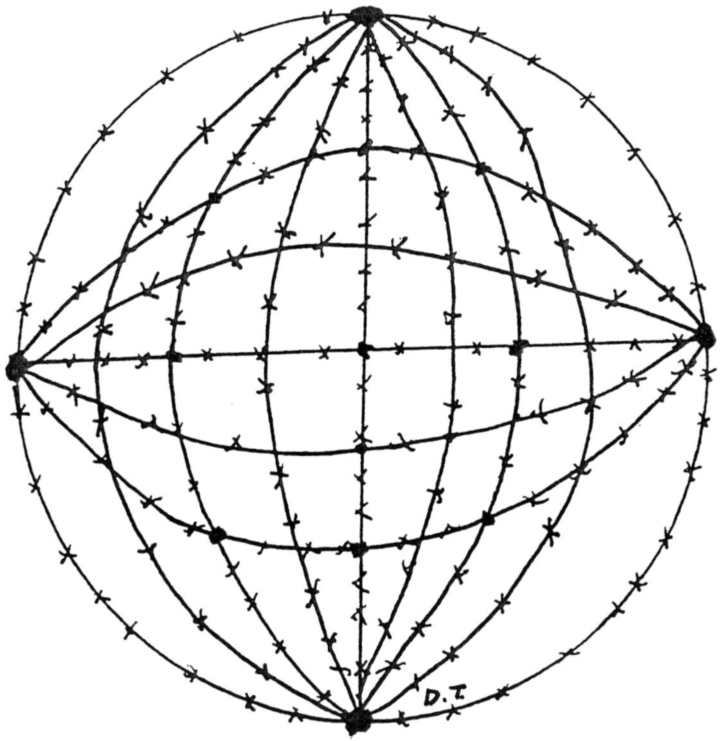

**GOOSEBERRIES**

**These round spheres are constructed on the spot by coiling wire and tying it together in whatever size needed by the troops. They were forced into windows, doorways, other entrances to trenches or fortifications and were effective in slowing passage.**

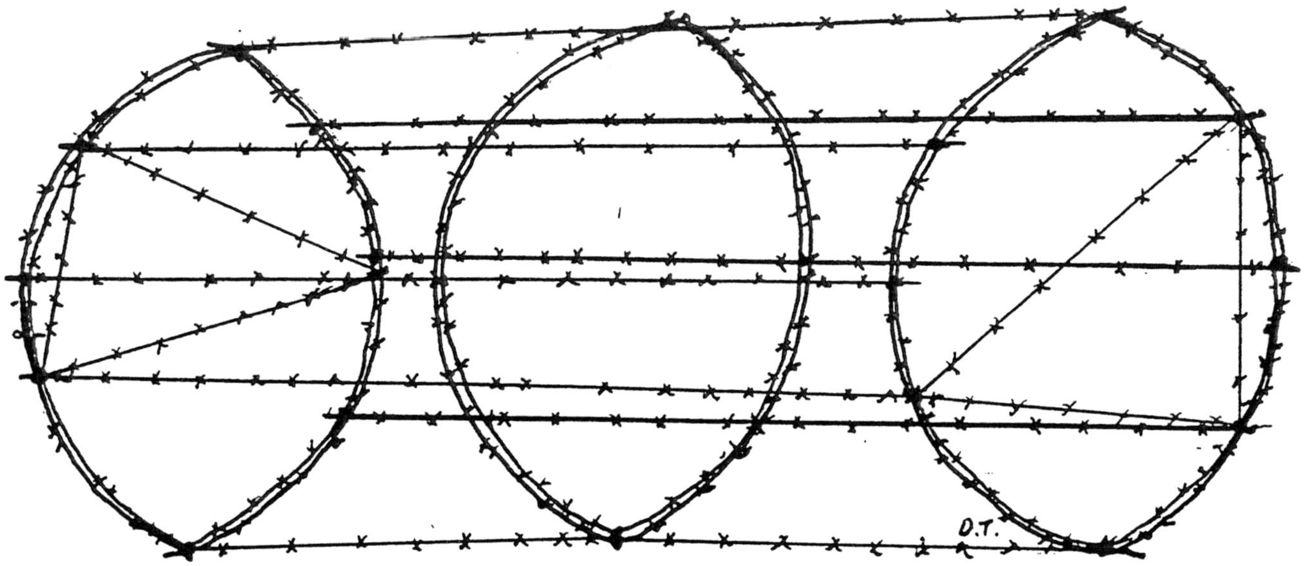

**RIBARD WIRE**

**These lightweight cylinders, constructed by troops on the spot, were easy to move yet were effective in controlling passage. The design was used where support posts could not be driven into hard or frozen surfaces. The cylinders could be fastened in lines, double lines, or double lines and stacked two or more high, all without support posts.**

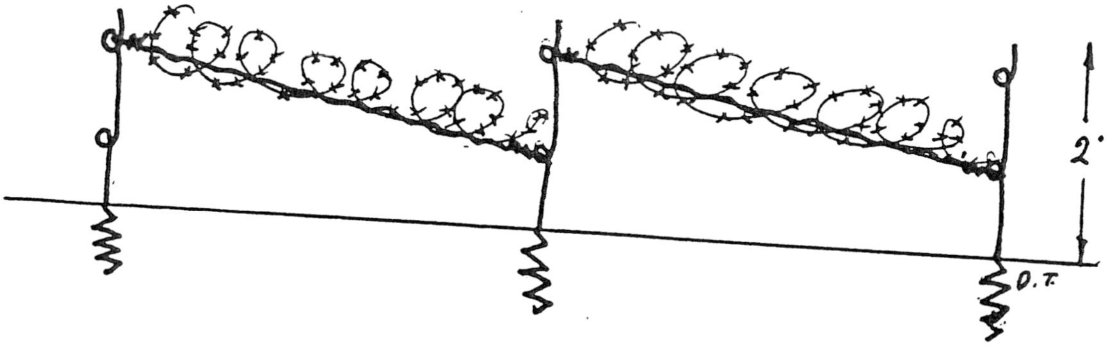


**END VIEW OF A STANDARD LOW WIRE ENTANGLEMENT**

**Each of the entanglement versions above were built in multiples both in depth and length depending on the terrain and fortification design. Each used the screw-eye type posts and stakes of various lengths.**

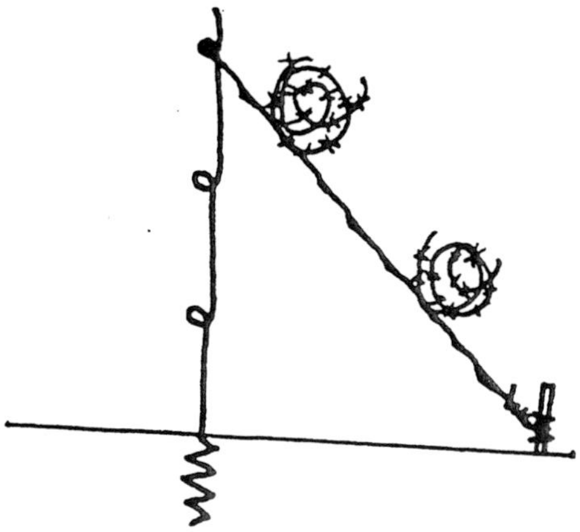

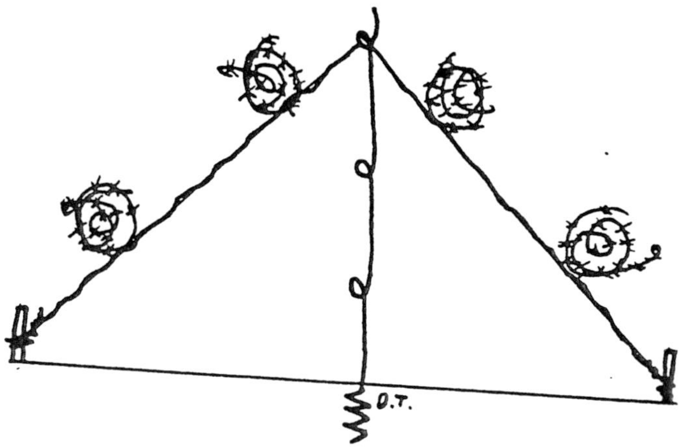


**END VIEW OF SINGLE APRON AND DOUBLE APRON FENCE**

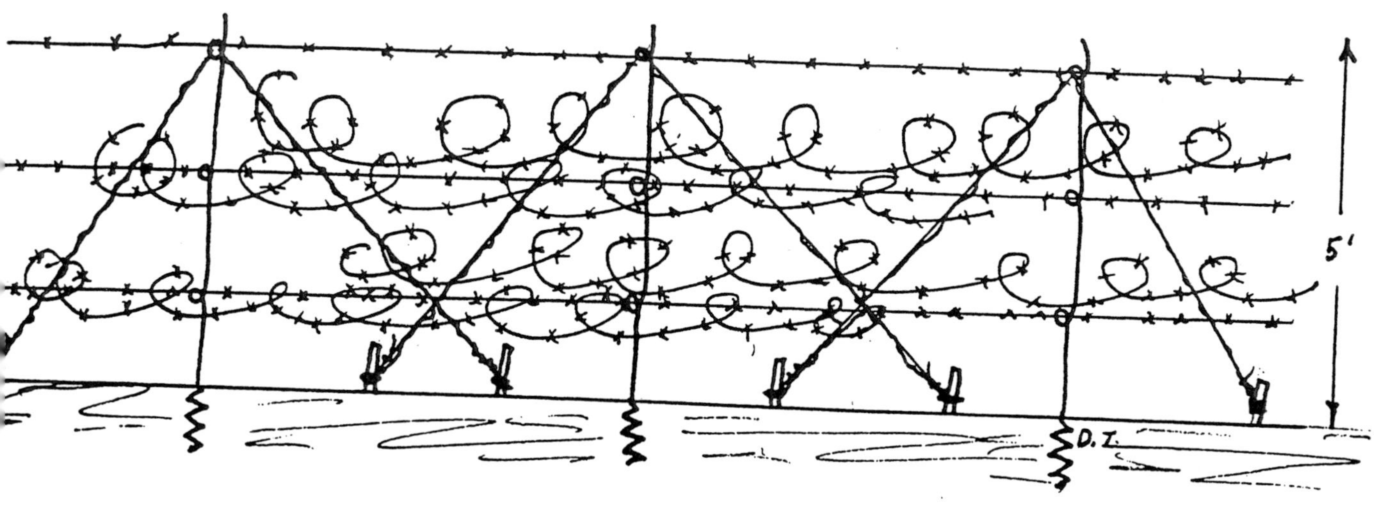


**STANDARD DOUBLE APRON FENCE**

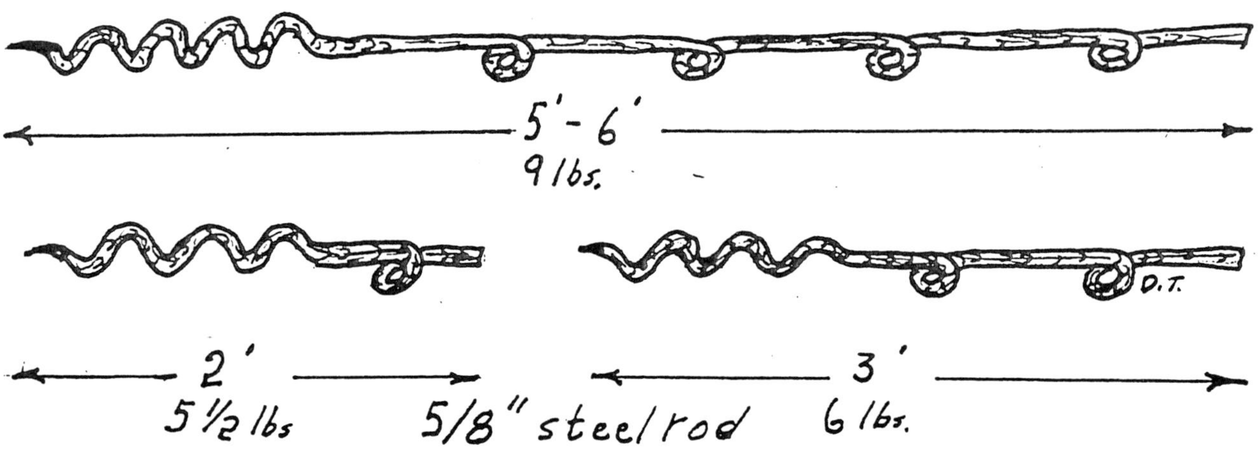


**STEEL SCREW-EYE POSTS**

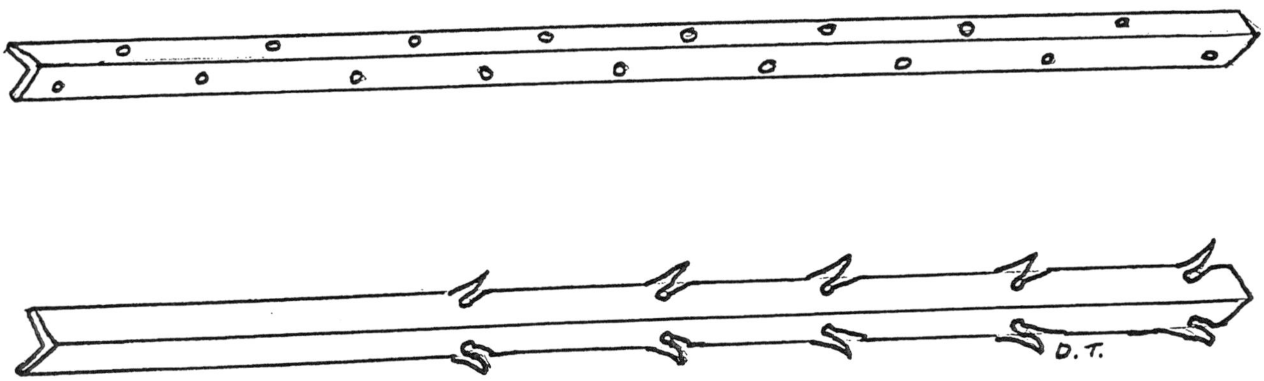


**ANGLE IRON POSTS**

**A FORREST OF ENTANGLEMENT WIRE- WWI**

CONVENTIONAL BARRIER FENCE-AUSTRIA WWI

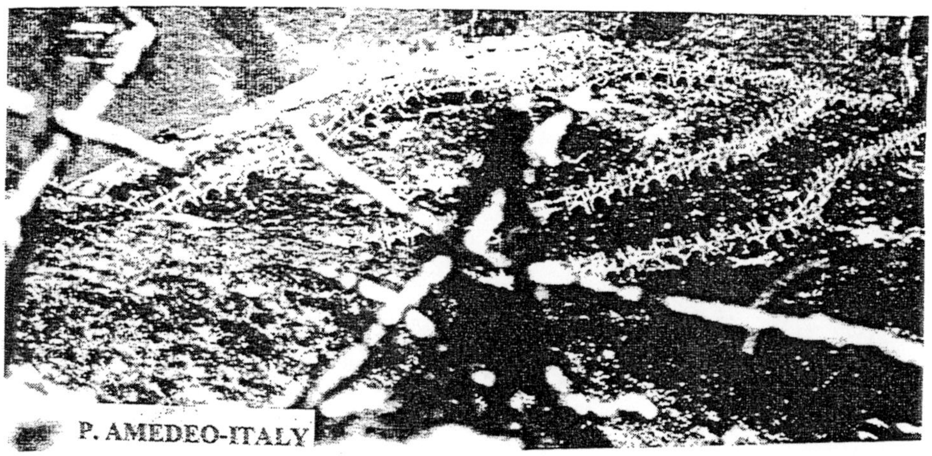


EXTREMELY LARGE BARBED CONCERTINA-WWI

AUSTRIAN BORDER-WWI-1915-1918

KNIFE REST DESI GN BARRIER-WWI

MORE KNIFE REST TYPE BARRIER-WWI

TRENCHES AND TERRACED BARRIERS-AUSTRIA WINTER 1917

SECTIONAL KNIFE REST BARRIERS-ITALY 1917

REAR GUARD AREA-FOLGARIA-1915

SAN GABRIELE MOUNTAIN-WWI-1915

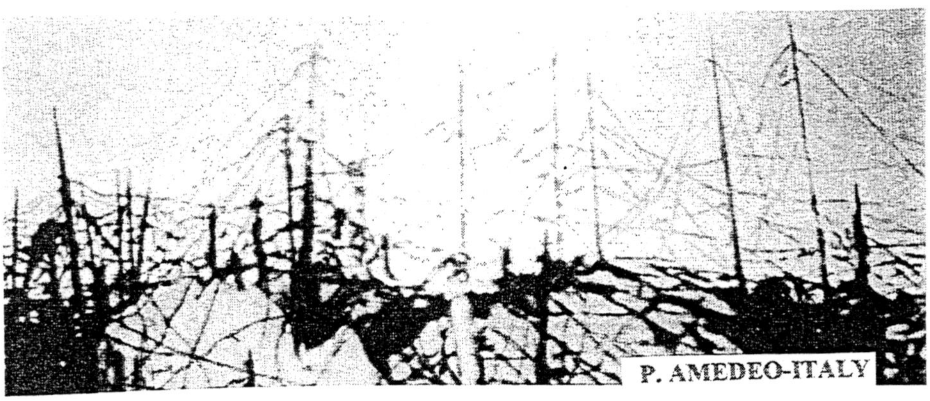


AUSTRIAN PROTECTIONS-FOLGARIA-1916

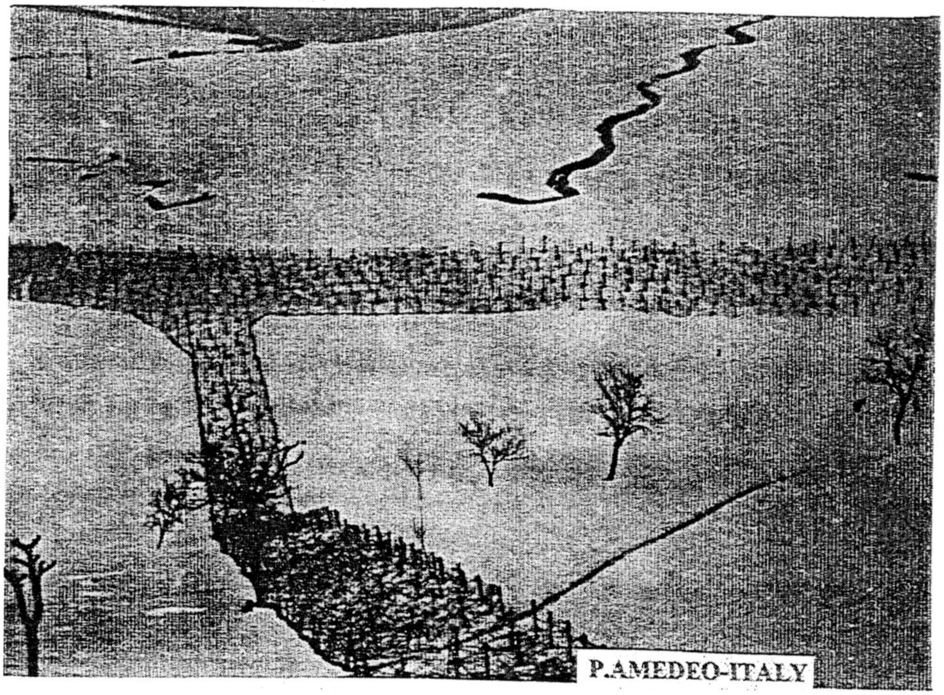


ITALIAN OBSTACLE-RIVER PIARE-WWI

NERVESIA NEAR RIVER PIARE-ITALY WWI

## SUGGESTIONS FOR FURTHER STUDY

Documents- Wire Entanglements Addenda No. 1 U. S. Army (1904-1918)
Military History Institute, Carlisle Barracks, Pensylvania

Entanglement Installation Books have been printed by the various branches of the military since 1904. They were updated year to year as needed. Any service connected library should have these.

Most of the books listed as suggested reading contain bits of information about wire barrier information.

Some information under "Beach Barriers" exists but was unavailable by mail request.

We believe an extensive search personally through the U.S. Army Military History Institute Library would reveal additional information.

# SECTION V.

## MANUFACTURING WARWIRE

**The story of The Lane Myers Company**
**of**
**Protection, Kansas**

**Makers of "Concertina Wire"**

## INTRODUCTION TO SECTION V.

## MANUFACTURING WARWIRE

It would be impossible to learn of the total number of companies throughout the world who, at one time or another, manufactured entanglement wire. Records of quantity of warwire produced are equally hard to find. Some of this can be blamed on the classified information restrictions. However, when the total production of warwire is compared to the total production of standard livestock type barbed wire, the military wire production is minimal. Making warwire was not attractive to large concerns.

Although the big steel companies were not interested in manufacturing warwire, they were interested in supplying the raw coils to small companies who were interested. Steel companies such as Colorado Fuel & Iron at Pueblo, Colorado made special efforts to help small companies using their raw products, bid on government contracts.

One of these small concerns is the Lane Myers Company of Protection, Kansas. Located in southern Kansas in a small farming community, the welding and machine shop was an ideal potential manufacturer of military products. With the help of CF&I, Lane Myers began to bid on the various contracts offered by the government.

The Lane Myers Company had been a customer of CF&I from the start. Steel products purchased from the steel company grew from five hundred tons per year to two thousand tons as the small company expanded.

In 1951, Myers was the low bidder on a $600,000 contract to provide drive-type steel posts for use in installing entanglement barriers. Actual government specifications are vague, but employees recall the posts were two inch angle iron about six feet long.

Some posts were drilled for wire attachment, others had "torn ear" type hooks along the edges to hang wire. The finished posts were tied in bundles light enough for a soldier to carry one bundle at a time during installation.

Myer's first "Concertina Wire" contract was awarded in 1954 with a $400,000 bid. The company had to "tool up" to make concertina wire. They built their own machines to make the wire and handle the packing and shipping chores.

This contract was notable as Myers beat out established manufacturing plants in Battle Creek, Michigan, Trenton, New Jersey, England and Belgium. After production started, and the government saw the Myers company could do the job, another $970,000 contract followed.

Production grew to a peak in 1966 when Myers produced $3.5 million dollars worth of entanglement wire. The company eventually became the number one producer of concertina wire for the U.S. Military, shipping the huge bales of warwire all over the world. During peak production periods, Myers used three eighteen ton truck loads of raw wire coils weekly delivered from CF&I in Pueblo.

The finished concertinas were shipped out in fifteen hundred pound bales, eight feet long and forty inches high. Each bale contained twenty-eight compressed concertina coils, capable of stretching into fifty-six lineal feet of barrier. It could also be re-rolled and reused if not damaged. The Lane Myers Company can be proud of their record in the manufacture of warwire.

Original office and main plant of the Lane Myers Manufacturing plant in Protection, Kansas.

# Protection from Protection

Lane Myers, the company president, checks the tough spring wire as it is delivered from mill. Tensile strength of the wire is rated at 200,000 p. s. i.

Machines used to make Concertina Barbed wire.

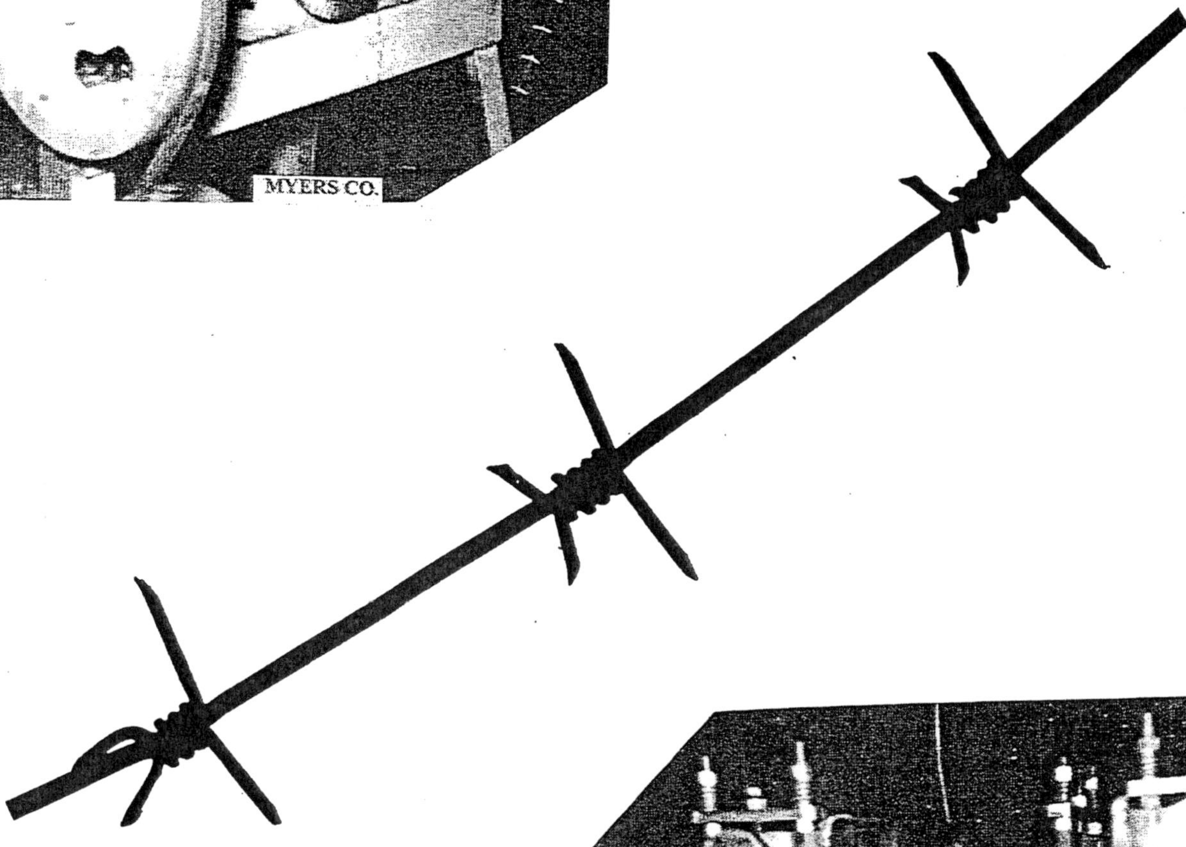

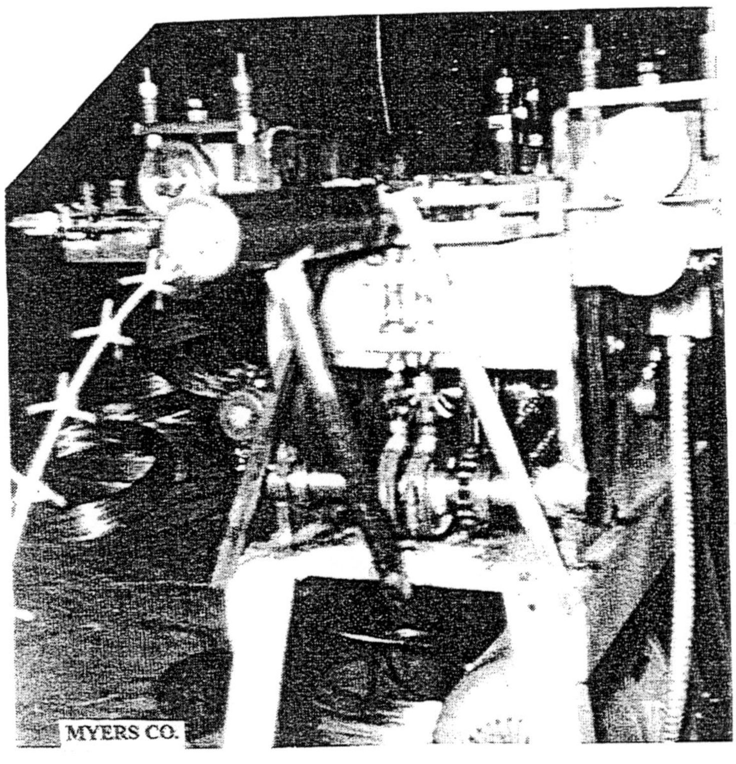


These machines ran 24 hours a day when a contract was secured.

**Fabricated concertina wire is fed from the machine onto reels. Five such machines are in Myers plant. Each wire machine produces 18,000 barbs per hour**

Single line Concertina wire on reel ready to be clipped in parallell.

Concertina coil clipped together ready to pressed down for tying.

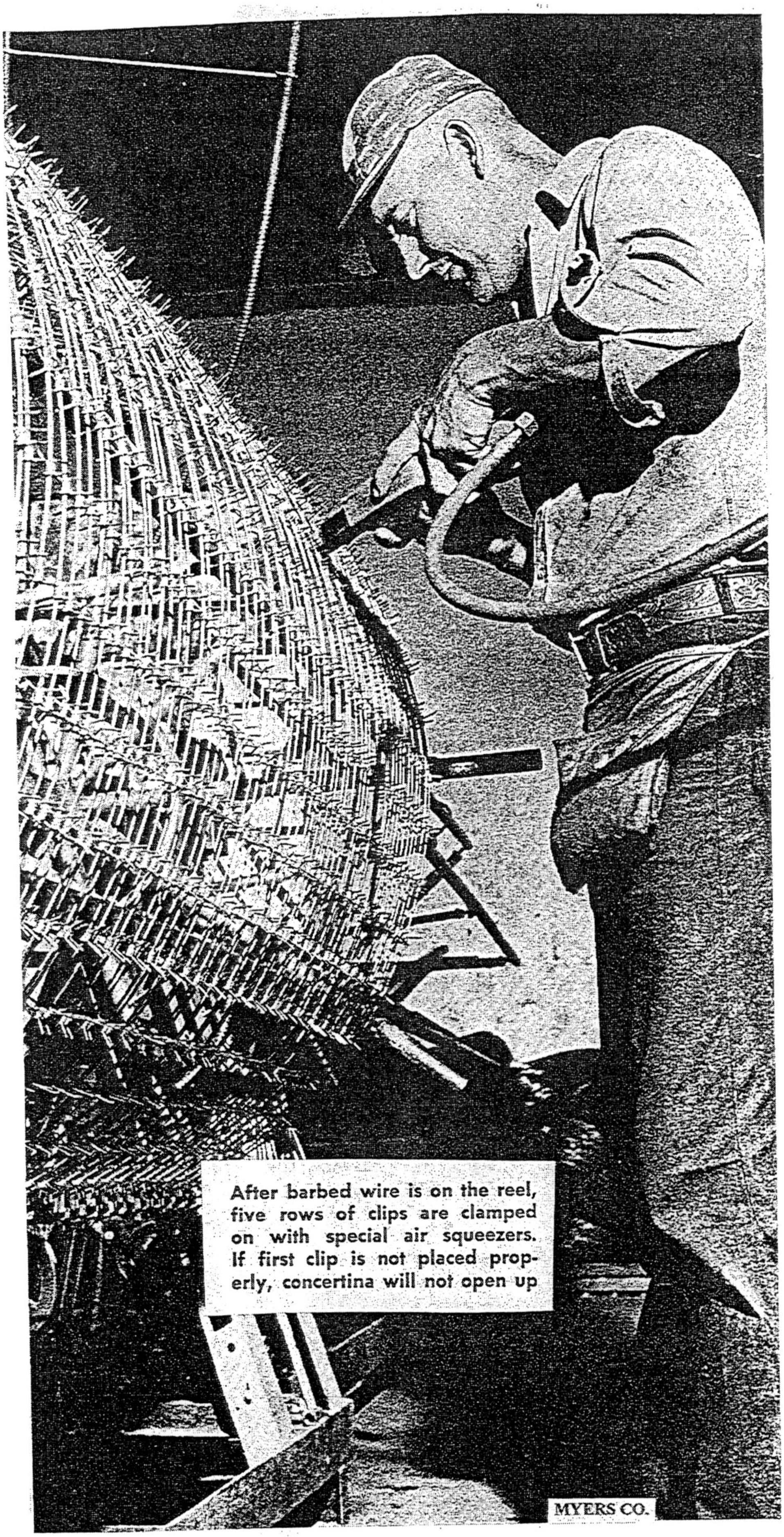


After barbed wire is on the reel, five rows of clips are clamped on with special air squeezers. If first clip is not placed properly, concertina will not open up

At the Lane Myers Company plant in Protection, Kansas two men are at work pressing a single unit bundle of "concertina" barb wire and tying it. Twenty-eight single units are included in the large bundle, which is approximately eight feet in length.

**After bundling, 28 units are compressed into a bale eight feet long for shipment. Bale is secured with four 2x6-inch boards. Units are shipped to U. S. Army engineer installations all over the world**

Close-up photo shows how bundles of "concertina" wire are loaded and stacked in gondolas for shipment to the Army. Lane Myers (standing on truck running board) personally supervises the operation.

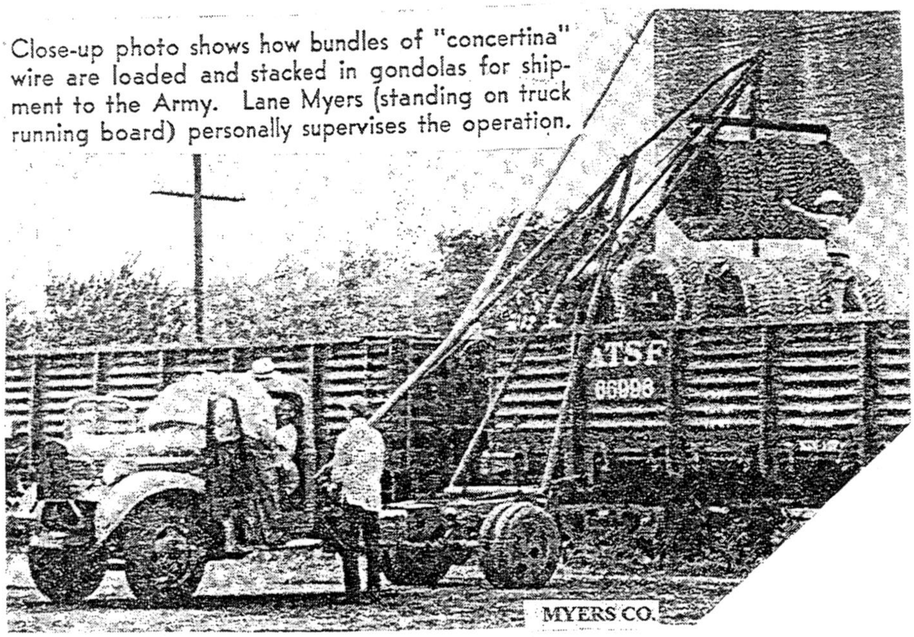


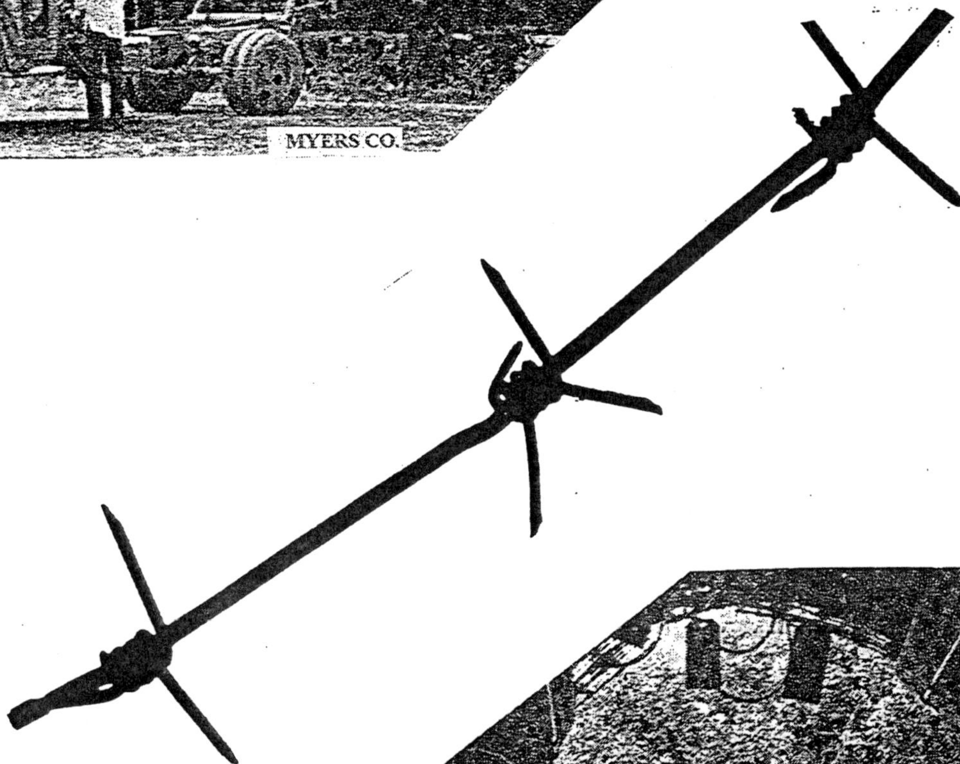

Picture taken from inside one of the bundles of "concertina" barb wire about to be shipped shows heavy boards used to hold units together. Standard length of bundle (8 ft.) makes perfect fit for boxcar.

Bales of concertinas await final inspection by government before shipment. Each bale contains 28 individual units

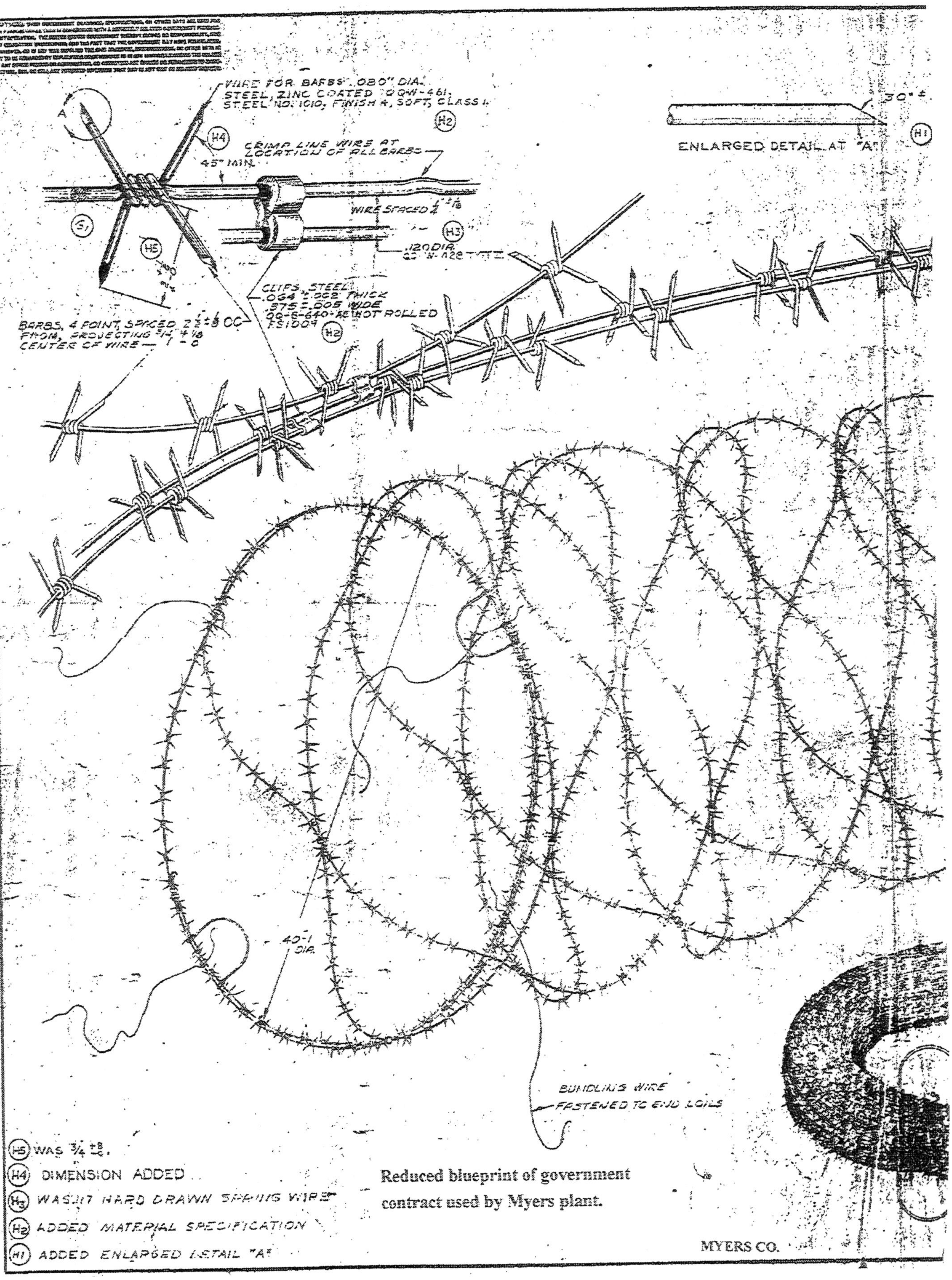


Reduced blueprint of government contract used by Myers plant.

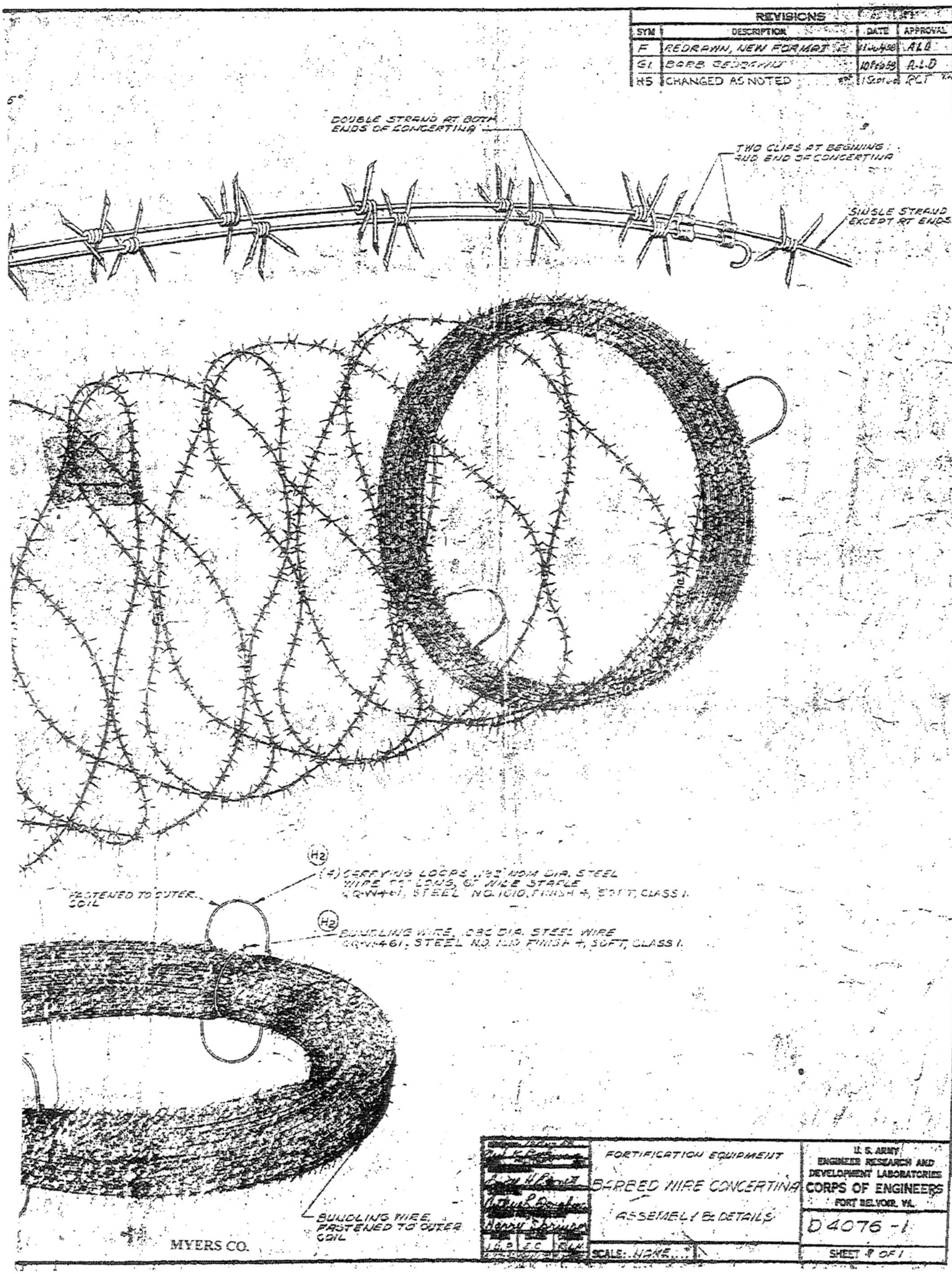

REVISIONS
SYM DESCRIPTION DATE APPROVAL
F REDRAWN, NEW FORMAT
G1 BARB
H5 CHANGED AS NOTED
DOUBLE STRAND AT BOTH ENDS OF CONCERTINA
TWO CLIPS AT BEGINING AND END OF CONCERTINA
SINGLE STRAND EXCEPT AT ENDS
FASTENED TO OUTER COIL
(4) CARRYING LOOPS
STEEL NO.1010, FINISH 4, SOFT, CLASS I.
BUNDLING WIRE
STEEL WIRE
FINISH 4, SOFT, CLASS I.
BUNDLING WIRE, FASTENED TO OUTER COIL
FORTIFICATION EQUIPMENT
BARBED WIRE CONCERTINA
ASSEMBLY & DETAILS
U. S. ARMY
ENGINEER RESEARCH AND DEVELOPMENT LABORATORIES
CORPS OF ENGINEERS
FORT BELVOIR, VA.
D 4076 -1
SCALE: NONE
SHEET 1 OF 1
MYERS CO.

## SUGGESTIONS FOR FURTHER STUDY

As all the Myers company files were loaned to us for study, we believe no futher information is available from this source.

Much of this type of information was contained in classified files and would be very hard to find. The Myers company is one of the few studied that are still in business today.

Several of the small manufacturing companies making special wires were of foreign origin and would be extremely hard to research.

# SECTION VI.

## WARWIRE PHOTOS

**A gallery of Warwire use.**

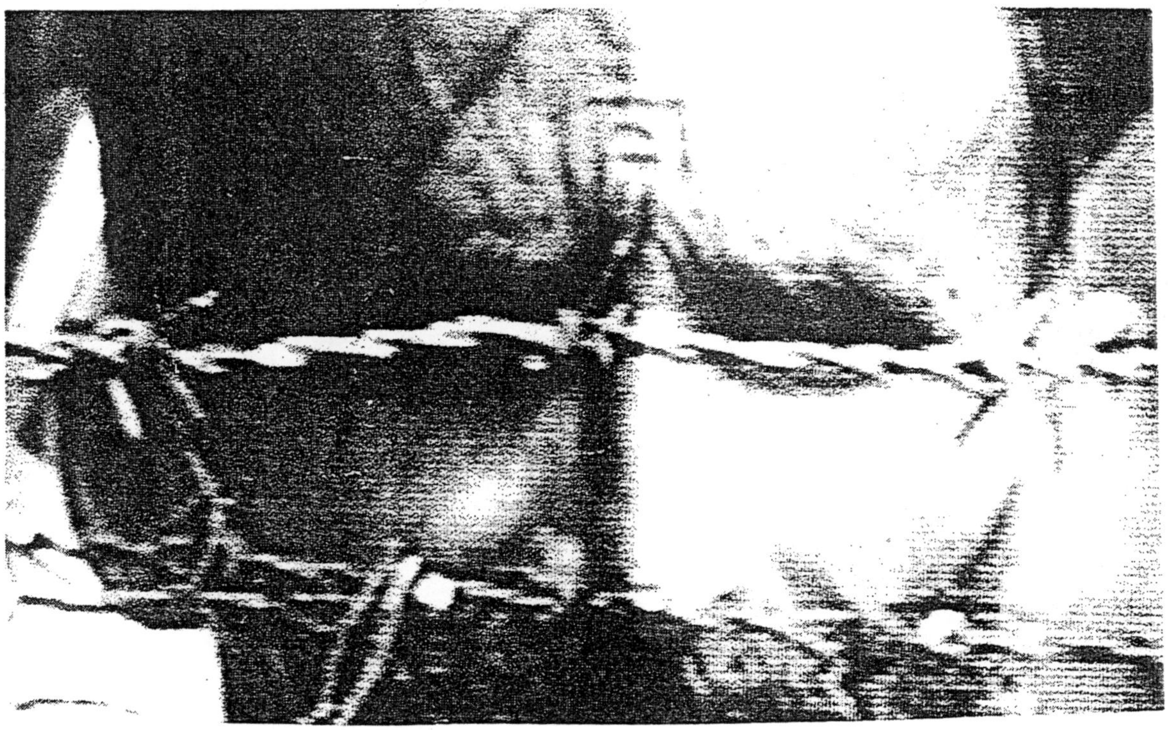

INTRODUCTION TO SECTION VI.

The photo gallery consists of small detailed portions of larger paintings and photographs. We use them only as graphic illustrations to reinforce history text, and for identification of entanglement wire and its related devices.

These illustrations are taken from the Devil's Rope Museum Library archives and are restricted to use for teaching aids and identification. All profits if any, from the sale of this book are dedicated to the operation of the museum and library and for the preservation of its artifacts.

We salute and thank those unknown artists, photographers, and historians who captured these moments in history.

**BARBED WIRE, TRENCHES, SAND BAGS AND GUARD DUTY- ITALY WWI**

Civil War painting detail showing telegraph wire at the bottom of a tree trunk and post barrier.

WWI painting detail showing trenches and early entanglement fencing.

WWI painting showing entanglement wire detail and resulting casualties.

WWI allied forces soldier cutting through a barbed wire entanglement.

WWI German soldier peering from trench through barbed wire barrier.

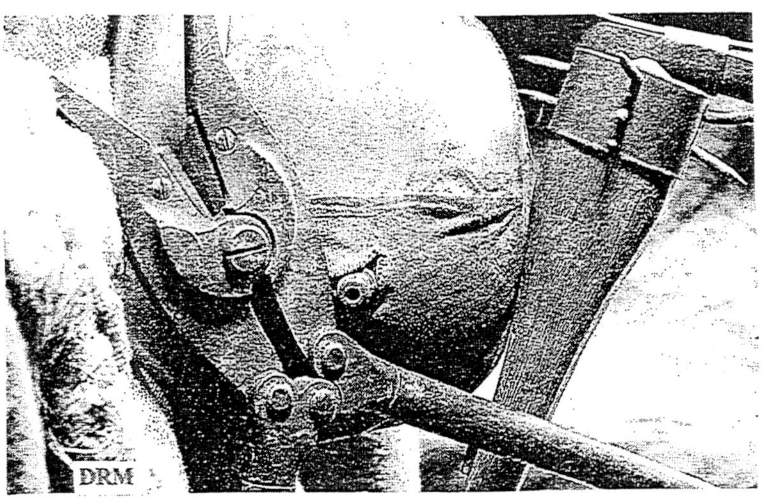


WWI bolt and wire cutters along with a hand made war club for close quarters fighting.

1890's - French/ German conflict, shows steel screw-eye posts piled in the trench awaiting installment in entanglement fencing.

1890's - troops in trenches ready for attack.

Boer War - painting detail showing screw-eye steel posts holding up barbed wire fencing.

Boer War - ribbon wire fences protecting block houses.

Boer War - roll of ribbon wire lying at a sabotaged railway crossing.

Boer War - English soldiers on guard behind barbed wire entanglements.

1905 Russo-Japanese War, painting detail showing Japanese troops chopping at ribbon wire entanglements with swords and hatchets.

1905 - Japanese troops using crude wire cutters on entanglement fencing.

1905 - Japanese troops climbing through smooth wire entanglements.

Different types and designs of beach barriers.

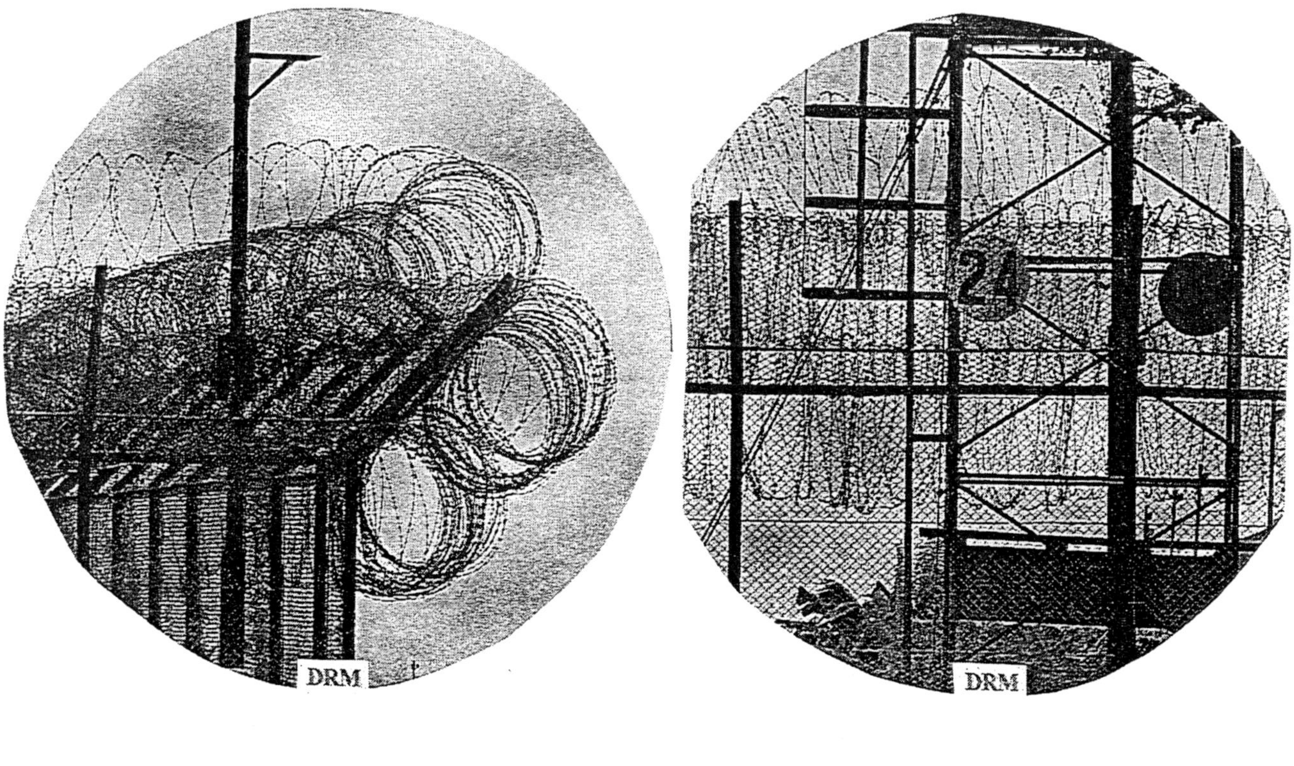


Details of Concertina coils of barbed wire used
in stockade construction.

WWII - Concertina wire being used in a temporary street barricade.

Persian Gulf War - U.S. soldier cutting razor wire with bolt cutters.

Iraq prisoners of war captured and being held behind Concertina coils of barbed wire.

Vietnam - entanglement fencing around a supply depot.

Vietnam - training exercises teaching soldiers to cope with entanglement barriers.

Vietnamese children being held away from a land mined area by Concertina coil fencing.

AUSTRIAN SOLDIERS ON PATROL
ITALIAN FRONT-1915-1918

ITALIAN ALPINE SENTRY AT TONALE PASS
1915-1918 WWI

**BARBED WIRE AT 3,890 METERS HIGH**
**MOUNTAIN PASS-ITALY-WWI**

**ADVERTISING PHOTO-AUSTRIAN SOLDIERS**

AUSTRIAN SOLDIERS ATTACKING ITALIAN
POSTS-WWI

CHARGE AT CARSO-WWI 1916

GERMAN SOLDIERS STACKING SPOOLS OF BARBED WIRE
IN A GUERNSEY STORAGE DEPOT- WWI

## SECTION VII.

## QUESTIONS AND SUGGESTIONS FOR COLLECTORS

## QUESTIONS, ANSWERS, AND SUGGESTIONS FOR THE COLLECTOR

This section is dedicated to anyone who has a collection already started, or to anyone who might be interested in starting a collection. We offer suggestions only and will discuss some of the things to consider as you begin to acumulate artifacts.

First, there is no mandatory, set in stone procedure you must follow. However, the experiences of veteran collectors will get you started on the right foot, and can save you time and money in your quest. Please keep in mind, these are suggestions "only."

Several obvious questions will be presented as well as answers based on experience. Other questions you may have can all be answered by asking other collectors or by researching your local library.

### What can I collect?

This must be answered by, "whatever interests you." Almost all great collections started with items chosen from a narrow field, and expanded as interest grew. As your knowledge grows, so does your interest.

An example of this is when a collector finds an interesting item. The search for the name and use reveals it is one of a group of items that are used to do something. This creates the urge to find the rest of the group.

Another example is when you learn the item was packed or carried in a special box or scabbard, or unique packet. This leads to a search for the carrier item.

We suggest you choose an area of interest, read and learn all you can about the subject, find and purchase an item, then clean, repair, and protect it from further deterioration. Your interest, efforts in the search, and personal care has increased the item's value, and your knowledge on the subject. You might already have a second item chosen for which to search.

### Once found, what is it?

The search for the name, use, age, and history of an item can be as facinating as the original hunt. This search can be done by asking veteran collectors about the object, asking old timers who may remember using the object, or searching the library resources.

As this information is acumulated, we suggest it be written down in some manner. We suggest you use a well made, sturdy notebook of some sort. Give the item a personal number, log the name, date found, and any other information you have gathered. In some manner, record the price and expense of your purchase. Also leave a column where you can pencil in the latest price noticed at an auction or antique store. This will keep you posted on any increase or decrease in price on the current market. As your collection grows, the logbook will prevent you from duplication, and mis-identification of items.

### How do I store it?

The answer here depends on the size and material makeup of the collection. Almost any dry, sturdy container will suffice. Excellent containers may be purchased at an economical price at Army Surplus stores. Moisture, sunlight, and handling are the major enemies of all artifacts. Paper items should always be stored in acid-free envelopes or sheet protectors. These can be purchased at any museum supply or most office supply houses.

Cloth and leather items require more and different types of care. Read up on this before you do it.

## How do I mount it for display?

Again this is up to the owner's discretion. Questions to consider might be, how much space do I have available? Is my spouse and family interested? Will my display fit in with the existing decor of my home or office? Will the display be permanent, or should I make it mobile?

Some study should go into mounting as it is both costly and time consuming. Small plaque type mountings could contain a collage of similiar type objects. Military subject plaques could contain identification patches, war medals or ribbons. Background could be natural finished wood, leather, or camouflage fabric.

If the items are to be displayed at museums, collector shows, etc., some thought should be given to protection during hauling and from theft.

A professional touch can be given to mountings by adding labels, a story or text telling what, when, and where about the items.

## What should I watch for in mounting?

Choose an attractive hardwood for framing or edging. Pick a background material that does not detract attention from the objects displayed. Choose a material that doesn't fade or collect dust easily. Attach the items firmly with wire or screws, or enclose the items with plex-iglass fronts. If possible, make the display free-standing or wall hanging. Make labels and text bold black print at least 3/4 inches high. If display is to be moved frequently, make a cardboard box for protection. Study other collector's mountings, museum mountings, and magazine photos of collections. If possible, choose a combination of mounting materials that can be duplicated later if other displays are added.

## Why should I make my collection mobile?

Experience by veteran collectors show that eventually all good collections have the opportunity to be displayed before the public. A mobile, protected display is easy to show. Another reason is a well mounted, attractive display adds to the value of the collection. When the attractive display is also mobile and protected, it adds even more value.

## How is the most common mobile display constructed?

We suggest the display size be 24 inches X 24 inches, or 24 inches X 48 inches, to prevent waste of construction materials and maintain a reasonable size for handling. The thickness or depth of each board is determined by the thickness of the objects to be mounted. We also suggest that two displays be placed face to face and hinged on one side and fastened on the opposite side so that when folded out and left at a slight angle, the combined display is free standing. This angled position allows stacking either on floors or tables. Small hangers on the back allows a wall hanging position also if desired. Comfortable, strong handles that are flexible or fold down flat should be attached opposite the hinged side.

These shadow box type display boards can be mortised on the top side to allow for the placement of plexiglass covers. We suggest using small metal type pan head screws to fasten down the plexiglass covers. Drill the screw holes larger than the screw shanks to allow for expansion and contraction of the plexiglass.

If wire is used to mount the items on the boards, flatten the twisted ends against the wooden surface to prevent snagging. Additional protection can be attained by dabbing a bit of hot glue over the wire twists. Very lightweight cases can be made from slab door skins found at some discount salvage stores. Cases of medium weight can be made of wood paneling. Heavy cases should be made from 1/4 inch plywood. Brass hinges, handles, and trunk type fasteners keep their lustre and function well. Clear varnish makes the outside of the cases look good and helps with keeping the surface clean. A small coded label near the handle tells the case contents without opening. Be aware that any display before the public, is subject to theft and vandalism. Protect them at all times.

## Tell me about collecting Warwires.

Barbed wire collectors have been active for over fifty years accumulating huge collections of barbed wire, related tools, paperwork, and history. "Warwires" are just a small unique segment of these large collections.

Early on, collectors found that an 18 inch length of wire was needed to show all the details such as twist and barb spacing. Most wire samples are sold in this length. To save space, some modern collectors use 4 to 6 inch wire lengths. Most souvenir plaques and mobile collections use the shorter lengths.

A historically correct display should show all the wires, and combinations of wires used in warfare. Since many of the wires are similiar in appearance and the public unable to distinguish between, souvenir type displays may display only those wires uniquely different.

## How can I locate and purchase Warwire samples?

Many antique stores stock limited supplies of collectible barbed wire. Some military surplus stores are becoming aware that entanglement wire is a new collectible. However, the best source is a barbed wire dealer who stocks all types of barbed wire and wire tools.

## How do I locate a dealer?

Barbed wire dealers are located all over the U.S. Most live in the mid-west area. There were thirteen two and three day shows in 1997, most held in the midwest. The largest dealers participate in these shows with numerous tables filled with barbed wire samples for sale.

All of these dealers also sell by mail. All known wires have an identification number and name. Wire Guide Values are copyrighted, updated every two months, and published in The Barbed Wire Collector Magazine. Wire identification books are available to help novice collectors in their choices. A letter, or a phone call, or a visit to a show is all it takes to get in touch with barbed wire people.

## What price should I pay for warwire?

Warwires are mostly modern variations of old patents or the newer razor wires. The price of these older wires should be $2.00 or less for each 18 inch piece. A wire in excellent condition is worth more than rusty wires. With WWII over 50 years old now, prices of all war items should increase in value.

Razor wire sales are sporadic with most 18 inch pieces selling in the $2-4.00 range. This wire is extremely dangerous to handle and requires gloves and heavy cutters. Use every caution when handling and mounting razor wires.

## What price should I pay for warwire related items?

There are few warwire related item collectors at this time. All barbed wire collectors tool collectors, and military collectors have some of these items featured. At this time, the most popular item would be bayonets with wire cutting ability.

Gun shows, tool shows, military catalogs, and antique stores are the best source for buying bayonets. Prices range from $20.00 to $400.00 depending on the age and type of bayonet found. Most wire cutter bayonets should sell below $100.00.

Many of the unique patented devices shown in the tool section of "Warwire" were never manufactured. Those that were made in quantity will be reasonable in price. Others will be considered rare and will be higher in price.

## Why should I collect warwire items?

We believe the collecting of warwire, related devices, photos and paper items are in the infant stage. At first, finding items will be difficult, but as demand grows the items will become available and will increase in price. Heretofore, information about warwire items was hard to find. We hope this book will arouse interest bringing forth more information from the past.

We encourage any collector with warwire items on display to focus attention on these items with good labeling and a story about the item. If several items are owned, put them together into a separate display with a story. This will call attention to the items.

## How can I help locate warwire items?

Start by telling the Warwire story to others. Ask gun show dealers if they have such items available. Ask antique store dealers for the items. When the public begins showing interest in certain items, the hunt will be on. Within a short time, warwire items will appear.

Give me some resources to contact about barbed wire.

1. A subscription to The Barbed Wire Collector Magazine costs $15.00 per year with six sixteen page issues mailed each year. The magazine contains phone numbers and addresses of the main dealers in the U.S. It also gives the upcoming wire show dates and locations. The Wire Value Guides are updated and published in each issue. Send $15 to The Barbed Wire Collector at Box 290, McLean, TX 79057. Phone 806-779-2225 or 806-779-3164.

2. There are two barbed wire museums in the U.S. The largest barbed wire museum in the world is located in McLean, TX at the address listed above. Over 14,000 sq. ft. of displays, library, souvenir shop, community room and gallery await the visitor. Summer hours are Tuesday -Saturday, 10:00 - 4:00, Sundays 1:00 - 4:00. Winter hours (November 1-March 1) Thursday -Saturday 10:00-4:00.

The Kansas Barbed Wire Museum is located in La Crosse, Kansas. Write to P.O. Box 717, La Crosse, KS 67548 for information on hours open.

3. For barbed wire development and early history, read The Wire That Fenced The West, by Henry and Frances McCallum, University of Oklahoma Press, 1965, ISBN:0-8061-1559-9.

4. For barbed wire identification, see The Bobbed Wire Bible IX, by Jack Glover, 1996, Cow Puddle Press. Another is Barbs, Prongs, Points, Prickers, and Stickers, by Robert T. Clifton, University of Oklahoma Press, 1970, ISBN:0-8061-0876-2.

5. For barbed wire patent information, read Barriers, by Robert O. Campbell and Vernon Allison, Western Profiles Publishing Company, 1986, ISBN:0-937231-00-2.

6. For futher information on the POW history, read Nazi Prisoners Of War In America, by Arnold Krammer, Scarborough House/Publishers, 1979, ISBN:0-8128-8561-9.

Why should I attend a wire show?

Barbed wire shows are similiar to an antique, gun and knife, or western memorabilia shows. You will find a lot of barbed wire offered for sale. You will meet dealers personally, and voice your needs. If they don't have it, they usually know who does. All of the books listed above as well as the BWC magazine are usually for sale. Many antique items of all nature will be offered for sale as well as wire. A visit to a wire show is a learning experience.